Environmental Science, Engineering and Technology Series

THE AMAZON GOLD RUSH AND ENVIRONMENTAL MERCURY CONTAMINATION

Environmental Science, Engineering and Technology Series

Nitrous Oxide Emissions Research Progress
Adam I. Sheldon and Edward P. Barnhart (Editors)
2009 ISBN: 978-1-60692-267-5

The Amazon Gold Rush and Environmental Mercury Contamination
Daniel Marcos Bonotto and Ene Glória da Silveira
2009 ISBN: 978-1-60741-609-8

Environmental Science, Engineering and Technology Series

The Amazon Gold Rush and Environmental Mercury Contamination

Daniel Marcos Bonotto
and
Ene Glória da Silveira

Nova Science Publishers, Inc.
New York

For permission to use material from this book please contact us:
Telephone 631-231-7269; Fax 631-231-8175
Web Site: http://www.novapublishers.com

Library of Congress Cataloging-in-Publication Data

ISBN: 978-1-60741-609-8

Available upon request

Published by Nova Science Publishers, Inc. ✝ New York

CONTENTS

About the Authors

Daniel Marcos Bonotto

Physicist, Assistant Research Professor of Geochemistry specialized in hydrogeochemical approaches utilizing natural radioelements as environmental tracers.

Departamento de Petrologia e Metalogenia
IGCE-Instituto de Geociências e Ciências Exatas
UNESP-Universidade Estadual Paulista "Júlio de Mesquita Filho"
Av. 24-A, No. 1515 – CP 178
13506-900-Rio Claro-SP
Brazil

Ene Glória da Silveira

Geologist, PhD in Regional Geology and researcher actively engaged in investigations focusing the presence of heavy metals, mainly mercury, in the Amazon area.

Departamento de Química
UNIR-Fundação Universidade Federal de Rondônia
Av. Presidente Dutra, No. 2965 – Centro
78900-500-Porto Velho-RO
Brazil

ACKNOWLEDGMENTS

CNPq (National Council for Scientific and Technologic Development), FAPESP (Foundation Supporting Research in São Paulo State), CAPES (Coordination for the Development of the Personal from Universities), UNESP (University of the State of São Paulo) and UNIR (Federal University of Rondônia Foundation) in Brazil are greatly thanked for financial and logistic supports of the main studies related to this book.

"... Owing to the method on which your science is based, it is not fit to discover the intimate bonds which link things together, and reveal their essence..."

The Great Synthesis
Pietro Ubaldi (1932 – 1935)

PREFACE

The importance of the Amazon area to sustain the global equilibrium in the environment has been recognized worldwide. This has been much more accentuated in the present days due to the intense debate related to global warming. Consequently, all initiatives/studies directed to a better knowledge/management of that huge environment are welcome and needed. This book is a contribution to this task, as gold has been exploited intensively in the Brazilian Amazon during the past 30 years using *garimpo* methods (small-scale gold mining), where the elemental mercury (Hg) used in amalgamating the gold, the final stage of the ore dressing process, has caused abnormal Hg concentrations in waterways. This has occurred in several areas of the Amazon region, where most of the ore prospected is alluvial. Particular attention to the Madeira River has been given since 1986 by several investigators. The main reason for this is that the Madeira River is the largest tributary of the Amazon River and the gold mining was officially allowed on a 350-km sector of the river, for its mid and upper reach, in the northwestern reach of the Amazon basin. Consequently, mercury was released from gold-mining fields to the atmosphere or to waterways in the metallic form, due to the large number of mechanical dredges operating simultaneously (about 6,000 during the peak mining activities). Although Hg^0 is relatively immobile in the aquatic environment and its solubility is low in water, Hg contamination in people living upstream and downstream from *garimpos* has been reported. The gold-mining activities on the Madeira River basin reduced substantially in the present days, i.e. it is practically absent. However, despite this, it is necessary a better understanding of the Hg behavior in tropical aquatic systems, mainly close to the most populated areas, as people may be still suffering toxicological consequences of the Hg releases in the past. Therefore, even in the present days, the knowledge of the mercury occurring in the aquatic system of the Madeira River basin is a great concern by local/international authorities and environmentalists, since it can contribute for identifying the effects of the anthropogenic Hg inputs relatively to the background reference levels expressing the natural Hg concentration. This book describes the results obtained on the analysis of samples of water, bottom sediments, suspended solids and fishes that were collected at the Madeira River basin, Brazil, with the purpose of investigating the mercury release in the aquatic environment as a consequence of the gold mining activities. The book is mainly directed to university students (undergraduate, graduate and postgraduate), researchers, decision makers, politicians, environmentalists, and interested public in general mainly engaged in the following subject areas: Geosciences, Ecology, Environmental Engineering, Chemistry, Biosciences, Hydrology, and Toxicology.

Chapter 1

THE MERCURY USE DURING THE AMAZON GOLD RUSH

Mercury is a liquid metal at room temperature that was known to the ancient Chinese and Hindus, and was found in Egyptian tombs that date from 1500 BC. In China, India and Tibet, mercury use was thought to prolong life, heal fractures, and maintain generally good health. China's first emperor, Qin Shi Huang Di was driven insane and killed by mercury pills intended to give him eternal life. It has been said that Qin Shi Huang Di was buried in a tomb that contained rivers of flowing mercury, representative of the rivers of China. The ancient Greeks used mercury in ointments and the Romans used it in cosmetics. By 500 BC mercury was used to make amalgams with other metals. The Indian word for alchemy is Rasavātam which means 'the way of mercury'. Alchemists often thought of mercury as the First Matter from which all metals were formed. Different metals could be produced by varying the quality and quantity of sulfur contained within the mercury. An ability to transform mercury into any metal resulted from the essentially mercurial quality of all metals. The purest of them was gold, and mercury was required for the transmutation of base (or impure) metals into gold as was the goal of many alchemists (Wikipedia, 2009a).

Hg is the modern chemical symbol for mercury. It comes from hydrargyrum, a latinized form of the Greek word `Yδραργυρος (hydrargyros), which is a compound word meaning 'water' and 'silver' - since it is liquid, like water, and yet has a silvery metallic sheen. The element was named after the Roman god Mercury, known for speed and mobility. It is associated with the planet Mercury. The astrological symbol for the planet is also one of the alchemical symbols for the metal. Mercury is the only metal for which the alchemical planetary name became the common name (Wikipedia, 2009a).

Mercury contamination is considered one of the worst hazards among the anthropogenic impacts upon the environment. Mercury is one of the few metal pollutants which has already caused human deaths due to ingestion of contaminated food. It is estimated that globally more than 800 humans have died and more than 17,000 have been afflicted due to mercury poisoning (D'Itri and D'Itri, 1977; Lacerda et al., 1995).

The diverse uses of mercury have resulted in severe contamination of fish in many regions with significant economic impacts on populations depending on fisheries and other aquatic resources. It was in the fishing villages of Minamata and Niigata in Japan, where in the 1950's and 1960's more than 100 people died and several hundreds were handicapped for life due to consumption of mercury contaminated fish (Lacerda et al., 1995).

However, the mercury toxicology started to be better investigated from that accident in Minamata, where one child with a nervous system dysfunction had entered into Shin Nihon Chisso Hospital in 21st April 1956. Afterwards, four another patients with similar symptoms had reached the Public Health Center of Kumamoto in 1st May, which has been considered the official discovery date of the Minamata Disease, i.e. a neurological syndrome caused by severe mercury poisoning. The symptoms include ataxia, numbness in the hands and feet, general muscle weakness, narrowing of the field of vision and damage to hearing and speech. In extreme cases, insanity paralysis, coma and death follow within weeks of the onset of symptoms (Wikipedia, 2009b).

One special committee was established in 1956 for investigating the illness that had affected 56 persons and whose origin was unknown until that date. The results of the investigation pointed out that the patients were living at Minamata bay and nearby and that their diet had mainly consisted in fishes and sea fruits. It was also identified the presence of organic mercury crystals in the wastes of the chemical industry Chisso. Mercury had been released into a river discharging in an oceanic area which was extensively utilized for acquiring food by small communities of the region. The marine fauna had been intoxicated and the highly toxic metal reached the human organs through food (Peazê, 2005). According to Lacerda et al. (1987) and Boaventura (1995), about 2200 intoxication cases and 887 deaths had occurred in the 1950's and 1960's, and 47 cases had been reported only in Niigata. Then, the seriousness of the situation has lead to an intense attention in most countries for the Hg presence in the environment. For instance, it was identified in Iraq 6530 intoxication cases and 459 deaths attributed to mercury in the 1970's (Peazê, 2005).

The illnesses and deaths related to the mercury contamination are typical examples of the high toxicity of the group of chemical elements named heavy metals. The release of industrial wastes is among the major contamination source in river waters by heavy metals. Metallurgic industries as well those related to the ink, chlorine and PVC plastic (vinyl) production have utilized mercury and others metals in their production lines, and, as a consequence, a part of the wastes has been discharged into the drainage. Others important sources of the environment contamination are the incinerators for urban and industrial litter, which cause their volatilization, forming ashes enriched in metals, mainly mercury, lead and cadmium (Peazê, 2005).

The heavy metals are highly reactive from the chemical point of view and this fact justifies the difficulty on finding them isolated in nature. They normally occur in very low concentrations and associated to others chemical elements, forming minerals in rocks. They can be absorbed by animal and vegetal tissues when released into water and, consequently, incorporated in the human food chain. The final fate of the metals inserted in the tissues of living organisms is the deposition in sediments, which are permanent reservoirs for the contamination of the aquatic flora and fauna. The heavy metals may also deposit in soil or water bodies of more distant regions due to the movement of the air masses (Peazê, 2005).

The majority of the living organisms only needs of some metals in very small amounts (micronutrients) like zinc, magnesium, cobalt and iron (constituents of the hemoglobin). These metals become toxic and dangerous to the human health when reach concentrations above some limits. On the other hand, lead, mercury, cadmium, chromium and arsenic neither naturally occur in any organism nor exert nutritional or biochemical functions in microorganisms, plants or animals, i.e. their presence in living organisms is harmful in any concentration. The production of these metals have increased since the discovery of the metallurgy by man and, additionally, their toxic effects generated permanent health problems to human beings and ecosystems (Peazê, 2005). The global mercury cycle results from extremely complex physical, chemical and biochemical processes involving volatilization, evaporation, transport and deposition, which can lead to the formation of methylmercury, its more toxic form, depending on the environmental conditions.

Thus, pollution by metals has a long history since it started with the extraction from ores in the Bronze Age. For instance, the estimated lead input (in tons per year) into the world atmosphere since 6000 B.P. indicates an increase of about 10^6 times (Salomons and Stigliani, 1995). Environmental problems caused by the utilization of Hg in gold and silver mining have been reported since Roman times. Circa 77 A.D. Roman authorities imported 5000 kg of Hg from Spain to be used in gold amalgamation in Italy (Lacerda et al., 1995). During colonial times between 1570 and 1900, it has been estimated that the Hg release into the environment from the extraction of silver and gold in central and northern South America reached 196,000 tons (Nriagu, 1993; Lacerda et al., 1995). During the late 1800's North America experienced its gold rush and the Hg related problems associated with it. Tailings containing Hg of up to 500 µg/g are still present in central Canada, while in the Carson River area over 100,000 tons of tailings containing Hg with concentrations up to 4,900 µg/g have been reworked, potentially mobilizing over 7,000 tons of Hg from tailings (Lechler and Miller, 1993; Lacerda et al., 1995).

The Amazon area spreads over 9 countries: Brazil, French Guiana, Suriname, Guyana, Venezuela, Colombia, Ecuador, Peru and Bolivia (Branco, 1989). Almost 67% of its surface area (~ 5 million km²) is inserted within the brazilian territory in the following states: Amazonas, Acre, Amapá, Pará, Rondônia, Roraima, Maranhão, Mato Grosso and Tocantins. On the other hand, only about 7.6% of the brazilian population is living there (IBGE, 2007). The surface area of the Amazon basin is about 7 million km², which is distributed over 6 countries and includes approximately 25,000 km of navigable rivers (Domínguez, 2004). Its borders are: west - Andes Cordillera, with typical elevations of 6,000 m; north – Guyanas Plateau, with mountain peaks of up to 3,000 m; south – Central Plateau, with typical altitudes of 1,200 m; east – Atlantic Ocean, to where all water is discharged (Brandini, 2005).

The Amazonas River starts flowing in the Andes Cordillera at Peru (5,600 m of altitude), intercepts a short part of south Colombia and flows through the northern portion of Brazil in the west-east direction until its discharge in the Atlantic Ocean between Amapá and Pará States. It receives the following names, along its 6.840 km of extension: Lloqueta, Apurimac, Ene, Tambo, Ucayali, Solimões and Amazonas. It is the largest in world either in extension or amount of water discharged (~ 200,000 m³/s during the rainy season) (Domínguez, 2004). The Peruvian plains are reached after an extension of 1,900 km and altitude difference of 5,440 m from the beginning of the flow. However, the altitude difference is only 60 m from the Peruvian plains up to the Atlantic Ocean. The total number of affluents corresponds to about 7,000 along its flow, where the principal in Brazil are: left margin - Içа, Japurá, Negro, Trombetas, Paru and Jari; right margin - Javari, Jataí, Juruá, Tefé, Madeira and Xingu. The discharge of Amazonas River is ~ 60 times higher than that of the Nile River, the second largest in world. The width of the Amazonas River is variable, ranging from 13 to 50 km in Manaus city (Amazonas State, Brazil), depending on the dry and rainy season. It is narrower (1,800 m) and deeper (50 m) at Óbidos city (Pará State in Brazil), where the discharge is

200,000 m³/s. The average water level high is 10 m, reaching 16 m in the rainy season. The amount of sediments released into Atlantic Ocean is huge, circa 800 billion tons per year (Brandini, 2005).

Thus, the Madeira River is one of the largest tributaries of the Amazon River (its total surface area corresponds to about 1.3×10^6 km^2). It extends from Bolívia and crosses the city of Porto Velho, capital of Rondônia State, which is the most populated site (population approximately 370,000, according to the 2007 census) along it. Its flow regime is dominated by the climate of the region as the Aw-tropical rainy climate (Köppen classification) characterizes Rondônia State, i.e. the relative air humidity ranges from 74.2 up to 90.8% and air temperature from 20.7 up to 32.2 °C (Nunes, 1999). The wet season (average rainfall = 250 mm/month) occurs between October and April, whereas the dry season (average rainfall = 20 mm/month) between May and September (Nunes, 1999).

In aquatic systems, heavy metals (including mercury) and other elements are mainly transported in the solid phase, either sorbed onto particle surfaces and coatings, or incorporated into mineral grains. Thus, 90-99 percent of the total metal load in rivers is transported in the particulate phase, depending on the geochemical behavior of the metal and the nature of the physical and chemical environment (Miller, 1997). Upon discharge to river waters, metal speciation or complexation, and other physical-chemical parameters such as pH, redox conditions, and nature of the solid-metal species interaction, determine the partitioning of the metal between solid and liquid phases, controlling its dispersion throughout the aquatic environment, where adsorption on fine particles has been recognized as a key process in its transport (Lockwood and Chen, 1973; Gambrel et al., 1980; Luoma, 1983).

The sources of trace metals in aquatic systems determine their distribution ratio between the aqueous and solid phases, when the metal contents of river solids depend significantly on allochthonous influences. Mining activities involve the extraction, processing, and disposal of large quantities of rock, being responsible, very often, by the release of metals that are introduced either directly or indirectly to rivers and streams (Salomons and Förstner, 1984). The major problem affecting mercury during mining activities is related to the fact that it is widely utilized as an amalgam in processing alluvial gold ores, mainly in rainforest areas (Pfeiffer et al., 1989; Cramer, 1990).

The gold exploitation in several areas of the Brazilian Amazon has caused environmental problems due to the presence of mercury in the biosphere, whether in liquid, ionic or vapor form (Martinelli et al., 1988; Veiga et al., 1991; Lacerda et al., 1991, 1995; Agaki et al., 1993; Hacon et al., 1995, 1997; Villas Bôas, 1997; Lacerda and Marins, 1997; Rodrigues Filho and Maddock, 1997; Roulet et al., 1998, 2000a, 2000b; Dolbec et al., 2000, 2001; Guimarães et al., 2000a, 2000b).

Special attention has been given to mercury inputs at the Madeira River in the Amazon area by numerous investigators (Maurice-Bourgoin et al., 1997, 1999, 2000; Lacerda, 1995; Malm et al., 1990, 1991, 1993; Pfeiffer et al., 1989, 1991; Malm, 1991; Lacerda et al., 1987, 1989; Pfeiffer and Lacerda, 1988) since is one of the largest tributary of Amazon River and the gold mining was officially allowed and practiced since the Andean headwaters, causing the release of metallic mercury to the atmosphere and waterways (Azcue et al., 1994).

The gold mining activities on the Madeira River extended from the Bolivian headwaters, on the Beni River, down to the city of Porto Velho, capital of Rondônia State. It became a highly significant activity in the 1980's, involving directly or indirectly about 50,000 persons.

Despite to be practically absent in the present days, mercury has been evaluated in the aquatic system by local authorities and environmentalists due to the possibility of occurring contamination in people (Cleary et al., 1994). Contamination of the aquatic food-chain of the Madeira River tributaries has been demonstrated by Maurice-Bourgoin et al. (2000), being reported that 86% of the piscivorous fishes collected in the Beni River exhibited high mercury concentrations, i.e. values exceeding by almost four times the WHO (1976) safety limit.

The mercury losses during the amalgam production in the Madeira River basin area is high as it has been estimated values released to rivers between 40 and 45% and to the atmosphere between 55 and 60% of the total initial amount utilized (Pfeiffer and Lacerda, 1988). However, more "pessimistic" values have been reported by CETEM (1991a) who considered that during the amalgam production had occurred volatilization to the atmosphere of 70% of the total amount of mercury utilized, which was not possible to be recovered; about 10% of this amount had been volatilized in the gold dealer shops, whereas 20% was lost in the amalgam wastes. Therefore, the intense mercury use during the amalgamation processes in the 1980's has constituted an important source of pollutants to the Amazonian environment, which can contaminate rivers, lakes, sediments, plants and fishes, with consequences to the human health.

The first systematic analyses of mercury for samples of soils, bottom sediments and particulate matter from rivers of the Rondônia State were realized by the Radioisotopes Laboratory of the Biophysics Institute "Carlos Chagas Filho" from *Universidade Federal do Rio de Janeiro* (UFRJ-Rio de Janeiro) and afterwards by the Environmental Geochemistry Laboratory from *Fundação Universidade Federal de Rondônia* (UNIR-Porto Velho). Most of the results here reported were described by Silveira (1998) as a sequence of the investigations performed by Pfeiffer and Lacerda (1988), Pfeiffer et al. (1989, 1990), Malm et al. (1989, 1990), among others. The mercury mobilization is specially focused in the sector of the Madeira River between Teotônio and Santo Antônio falls, close to Porto Velho city, as this site is the most populated along it. Data on the presence of mercury and others elements in waters, bottom sediments, particulate matter and fishes have been considered for trying to properly evaluate the scenario in the area investigated. Despite it may happens the absence of major information in some circumstances, the importance of the Amazonian area as well of the hydrological resources for the world in this century justifies such an effort for a better understanding of the mercury occurrence there.

Chapter 2

MERCURY IN THE ENVIRONMENT

The specialized literature about mercury is significant in Brazil and worldwide, mainly after the accident in Minamata and Niigata, Japan, where in the 1950's and 1960's more than 100 people died and several hundreds were handicapped for life due to consumption of mercury contaminated fish. This chapter focuses some aspects related to the presence of mercury in nature, which are important to understand the behavior of this element in waters and sediments.

SOME PROPERTIES OF MERCURY

Mercury is a high specific weight metal, shiny, silver colored and liquid at room temperature. Its atomic number is 80, belonging to the transition elements group (*d* block of the Periodic table). It has two paired electrons in *s* orbital and the last orbital (*d*) is filled (10 electrons). Its electronic configuration corresponds to K = 2, L = 8, M = 18, N = 32, O = 18 and P = 2, i.e. $1s^2\ 2s^2\ 2p^6\ 3s^2\ 3p^6\ 4s^2\ 3d^{10}\ 4p^6\ 5s^2\ 4d^{10}\ 5p^6\ 6s^2\ 4f^{14}\ 5d^{10}$.

Mercury has three oxidation states: 0, 1 and 2. The Hg^0 is the metallic mercury. It forms two bindings, as Hg^{+1} or Hg(I), like in the mercurous chloride (Cl-Hg-Hg-Cl). Valences higher than 2 are not found, because the removal of more electrons would destroy the symmetry of the filled *d* orbital ($5d^{10}$); however, several bivalent complexes are formed when two electrons are utilized in the bindings (Nehmi, 1995).

Some physical parameters related to mercury are: atomic weight = 200.59, melting point = -38.9 °C, boiling point = 356.9 °C, density at 20 °C = 13.5 $g.cm^{-3}$, vapor pressure at 20 °C = 1.2 μmHg, specific heat at 25 °C = 0.0331 cal/g, surface tension at 20 °C = 465 dyn/cm and viscosity at 20 °C = 1.55cp (USBM, 1970).

Mercury forms alloys with several metals like sodium (liquid amalgam when the Na content is lower than 1.5%), potassium, silver and gold ($AuHg_3$, Au_2Hg e Au_3Hg), as well with zinc and copper when bronze is formed or during the amalgam production with cadmium, tin and lead. Mercury is soluble with phosphorus and the color of the solution is clear (Lee, 1971).

Mercury nitrate [$Hg_2(NO_3)_2$] may be obtained by dissolution of metallic mercury in an excess of nitric acid; crystals needle-shaped without a well defined colour are formed after evaporation, whose formula is $Hg_2(NO_3)_2.2H_2O$. The compound is very soluble in aqueous solutions, and cannot be considered a complex, despite the formation of double nitrates.

Mercury sulfate (Hg_2SO_4) occurs in the crystalline form as a monoclinic prism. It is not colored and originates from solutions of mercury nitrate and diluted sulfuric acid. It is a little soluble in aqueous solutions, however, forms yellow greenish basic salts during its gradual decomposition by water, which become dark in the light presence.

THE ION HG_2^{2+}

Few mercury compounds are known in the monovalent form as they don't contain Hg^+ but the ion $(Hg\text{-}Hg)^{2+}$ where two Hg atoms are bonded through *6s* orbitals and the ion Hg_2^{2+} is formed from reduction of mercuric salts in aqueous solutions. In this type of bonding, one valence electron from each atom is shared by two atoms, while the others two valence electrons, one from each atom, are removed. Beyond the mercury, zinc and cadmium also form dinuclear unstable metallic ions like $(Zn\text{-}Zn)^{2+}$ and $(Cd\text{-}Cd)^{2+}$ that only are spectroscopically detected in mixtures of $Zn/ZnCl_2$ and $Cd/CdCl_2$ (Lee, 1996; Greenwood and Earnshaw, 1991).

The ion Hg_2^{2+} can be formed when metallic mercury is added to a silver nitrate solution; in a few seconds it will occur reduction of the ion Ag^+ and dissolution of the metallic mercury according to the reaction:

$$2Hg + 2Ag^+ \rightleftarrows Hg_2^{+2} + 2Ag \qquad \text{(amalgam)}$$

Measurements of the electromotive force have been performed in an electrolytic cell containing mercury nitrate at 25°C. The measured potential corresponded to 0.029V. Lee (1971, 1996) and Greenwood and Earnshaw (1991) utilized the Nernst equation and concluded that the number of electrons transferred in the reaction is 2, therefore, characterizing the presence of the ion Hg_2^{2+}.

The electric conductivity of mercury nitrate solutions also favors the formation of the ion Hg_2^{2+}, since does not occur the ionization reaction like that involving silver nitrate, i.e. $HgHNO_3 \rightarrow Hg^+ + NO_3^-$ because, on the other hand, it is verified a similar process to that affecting lead nitrate, as described by Lee (1971):

$$Pb(NO_3)_2 \rightarrow Pb^{2+} + 2NO_3^-$$
$$Hg_2(NO_3)_2 \rightarrow (Hg_2)^{2+} + 2NO_3^-$$

Raman spectroscopic analyses have indicated that mercury nitrate exhibits an extra line in the spectrum, attributed to the bonding [Hg-Hg] (Lee 1971, 1996). The ion Hg_2^{2+} posses all paired electrons that are diamagnetic like the great majority of the compounds, either in the solid or liquid state (Greenwood and Earnshaw, 1991). Cryoscopic measurements realized by Lee (1996) have suggested the occurrence of mercury nitrate dissolution into one Hg_2^{2+} ion and two NO_3^- ions, rather than one Hg^+ ion and one NO_3^- ion.

According to Cotton and Wilkinson (1982), the compounds $Hg_3(AlCl_4)_2$ and $Hg_4(AsF_6)_2$ contain three or four mercury atoms bonded themselves in a quasi linear way. These compounds may be generated from the following reactions:

$$2Hg + HgCl_2 + 2AlCl_3 \rightarrow Hg_3(AlCl_4)_2$$

$$4Hg + 3AsF_5 \rightarrow Hg_4(AsF_6)_2 + AsF_3$$

The crystalline structures of these compounds denote that the distance of the Hg atoms varies between 2.55 and 2.70 Å. The two Hg atoms in Hg_2^{2+} ion are bonded themselves by *6s* orbital that is a concept not possible to be extended for three or four atoms.

Mercury Halides

In the group of the halogenic elements, the fusion and boiling points increase according to the atomic number and, therefore, fluorine and chlorine are gaseous elements, whereas bromine is liquid and iodine is solid. All they are diatomic and, consequently, the bonding energy of the fluorine molecule is low (about 38 kcal/mol) as a consequence of the electronic repulsion. This causes these elements to react easily with metals, and mainly with mercury, forming the following halides: mercury chloride (Hg_2Cl_2), mercury iodide (Hg_2I_2), mercury bromide (Hg_2Br_2) and mercury fluoride (Hg_2F_2). The last halide is the most soluble, whereas the solubility of the others decreases according to the raising of the atomic number, as shown in table 1.

The mercury fluoride may be obtained from dissolution of mercury carbonate in hydrofluoric acid. The precipitated formed is yellow, becoming pale in the presence of light. The mercurous chloride may be obtained from Hg_2Cl_2 reduction in the presence of SO_4^{2-} or by $Hg_2(NO_3)_2$ treatment with hydrochloric acid. The product is calomel that is dark colored not only when exposed to light but also after treatment with ammonia, according to the reaction:

$$Hg_2Cl_2 + 2NH_3 \rightarrow ClHgNH_2 + Hg + NH_4Cl$$

The mercurous chloride is soluble in several organic solvents and has density lower than that expected by the formula Hg_2Cl_2 as a consequence of the decomposition of Hg_2Cl_2 in $HgCl_2$ and Hg. In aqueous solutions, it is verified the hydrolysis of $HgCl_2$, according to the reaction:

$$HgCl_2 + H_2O \rightleftarrows Hg(OH)Cl + H^+ + Cl^-$$

Table 1. Solubility and solubility product of the mercury halides. According to Greenwood and Earnshaw (1991)

Halide	Solubility at 25°C, mol/liter	Solubility Product
Hg_2Cl_2	1.03×10^{-6}	1.1×10^{-18}
Hg_2Br_2	3.7×10^{-8}	1.3×10^{-22}
Hg_2I_2	3.7×10^{-10}	4.5×10^{-29}

The mercury bromide (I) precipitates by KBr addition to mercury nitrate (I) solution; the potassium bromide precipitates forming a tetragonal laminar net, with vapor density corresponding to half of the formula Hg_2Br_2.

The mercury iodide (I) forms a greenish precipitate by addition of alkaline iodide into a mercury (I) solution and sometimes a yellow-greenish precipitate through mercury friction with iodine in the presence of ethanol. Its crystalline structure is similar to that of Hg_2Cl_2 and during heating in an aqueous solution there is a slow change in HgI_2 and, finally, in Hg, according to the equation:

$$Hg_2I_2 + 2I^- \rightarrow [HgI_4]^{-2} + Hg$$

The investigation of the crystalline structure of the mercury halides by X-ray diffratometry has demonstrated that the unitary cell does not contain individual ions, but linear molecules of the type X-Hg-Hg-X, for instance, [Hg_2Cl_2(Cl-Hg-Hg-Cl)]; it has also been demonstrated that the Hg-Hg distance varies from 2.5 to 2.7 Å, depending on the associated anions (Cotton and Wilkinson, 1982).

Mercury Salts

Hg(I) mercury salts may be obtained from Hg(II) salts, metallic mercury and dissolution of Hg(I) carbonate in acids. The Hg(I) salts are less soluble than Hg(II) salts, whose precipitation can produce them, forming stable complexes slightly ionized (Cotton and Wilkinson, 1982). The more soluble Hg(I) salts are the monoprotic oxiacids strongly associated to nitrates, chlorides and perchlorate that are highly ionized when in solution.

The standard potentials involving the three oxidation sates of mercury are defined by the reactions:

$$Hg_2^{2+} + 2e^- \rightleftarrows 2Hg \qquad \varepsilon^o = -\ 0.789\ V$$

$$2Hg^{2+} + 2e^- \rightleftarrows Hg_2^{2+} \qquad \varepsilon^o = -\ 0.920\ V$$

$$Hg^{2+} + 2e^- \rightleftarrows Hg^0 \qquad \varepsilon^o = -\ 0.854\ V$$

The oxidizing agents required to oxidize the ion Hg_2^{2+} to Hg^{2+} are normally strong and reducing agents like tin, zinc and hypophosphorous acid can reduce the ion Hg_2^{2+} to metallic mercury. According to the standard potentials values, the mercury salts can oxidize mercury to Hg(I), but not to Hg(II), as well Hg can reduce easily Hg^{2+} to Hg_2^{2+}, as demonstrates the reaction:

$$Hg_2^{2+} = Hg + Hg^{2+} \qquad \varepsilon^o = -0.131\ V$$

The equilibrium constant of this reaction corresponds to:

$$K = [Hg^{2+}] / [Hg_2^{2+}] = 6 \times 10^{-3}$$

which is subjected to disproportionation in the case of reduction in the Hg^{2+} activity (by precipitation or complexation) on a greater extent than that verified for the Hg_2^{2+} activity. Because there are several responsible reagents, the number of Hg(I) stable compounds is small.

When OH^- is added into a Hg_2^{2+} solution, it is formed the Hg precipitate, the Hg(I) oxide [HgO] and the Hg(I) hydroxide [$Hg_2(OH)_2$]. One typical reaction illustrating the process is:

$$Hg_2^{2+} + 2OH^- \rightarrow Hg + HgO(s) + H_2O$$

If it was possible to isolate the compounds, then, one strong base would be produced. For instance, a very insoluble mixture of Hg and HgS would be obtained with sulphide addition (Benes and Havlík, 1979), like represented by the following reaction:

$$Hg_2^{2+} + S^{2-} \rightarrow Hg + HgS(s)$$

Despite soluble, the mercury cyanide has a slight dissociation trend; it may be formed by the following reaction:

$$Hg_2^{2+} + 2CN^- \rightleftarrows Hg + Hg(CN)_2(aq)$$

Hosobara et al. (1961) have investigated the reaction of the Hg_2^{2+} ion with water at 25°C in a 0.5 M perchlorate solution and concluded that the water acidity of the water containing Hg(I) is due to its presence, as illustrates the reaction:

$$Hg_2^{2+} + H_2O \rightleftarrows Hg_2OH^+ + H^+$$

MERCURY MINERALS

Mercury has been often considered a rare element in nature because is liquid at the room temperature. Despite being a heavy metal, there are several volatile mercury forms, and cinnabar (HgS, Hg = 86% and S = 14%) is the major mercury mineral. The cinnabar is frequently impure and is associated to clays, iron oxides and bitumen. The mercury may be extracted from HgS by heating it in a current of air and condensing the vapor, according to the reaction:

$$HgS + O_2 \rightarrow Hg + SO_2$$

The cinnabar decomposition occurs at about 500°C, forming sulfur dioxide and mercury in the vapor state. The cinnabar occurrence is like hexagonal and romboedric often geminate crystals (Dana, 1976). Mercury also occurs in lower proportions in corderoite ($Hg_3S_2Cl_2$) and calomel (Hg_2Cl_2, Hg = 75%), among other minerals. Table 2 summarizes some of the major types of mercury minerals.

Cinnabar is the most important mercury mineral, whereas metacinnabar is the second one. The cinnabar and metacinnabar deposits have the same origins from aqueous solutions.

Dickson and Tunell (1958) determined the saturation curves for cinnabar and metacinnabar in the system $HgS-Na_2S-H_2O$ at 25°C, 50 °C and 75°C temperature and 1 atm pressure, as well the cinnabar solubility in the same system at 250°C temperature and 1.8 bar pressure, verifying that metacinnabar is more soluble than cinnabar. Schwarzenbach and Wildmer (1963) determined dissolved Hg content values corresponding to 5 ppb at 0.2 atm H_2S partial pressure, according to the following reaction:

$$HgS + 2H_2S\ (aq) \rightleftharpoons HgS\ (H_2S)_2$$

where $K = 10^{-4.25}$ at 20°C.

Some studies held by Barner and Rosemberger (1963) with alkaline solutions have demonstrated that the HgS solubility (black metacinnabar and red cinnabar) increases with the gradational pH raising that varies from 6 to 9 in the complexes forming metacinnabar at 20°C.

Table 2. Some major types of mercury minerals. According to Bethke (1956), Manasse (1921) and Mitra (1986)

MINERAL	FORMULA	SIMMETRY
Paraschachnerite	Ag_3Hg_2	Orthorhombic - Dipyramidal
Arquerite	HgAg	Cubic
Calomel	Hg_2Cl_2	Tetragonal
Cinnabar	HgS	Trigonal
Coloradoite	HgTe	Cubic
Corderoite	$Hg_3S_2Cl_2$	Isometric - tetartoidal
Eglestonite	Hg_4OCl_2	Cubic
Guadalcazarite	(Hg,Zn)S	Cubic
Hermesite	$Cu_2Hg_3Sb_2S_3$	Cubic
Kleinite	$Hg_2N(Cl.SO_4).xH_2O$	Hexagonal
Leviglianite	(Hg,Zn,Fe)S	Cubic
Livingstonite	$HgSb_4S_8$	Monoclinic
Magnolite	Hg_2TeO_4	Orthorhombic
Metacinnabar	$HgS-(HgS)_{80}(HgSe)_{20}$	Cubic
Montroydite	HgO	Orthorhombic
Moschellandsbergite	Ag_2Hg_3	Cubic
Mosesite	$Hg_2N(Cl,SO_4,MoO_4,CO_3).H_2O$	Cubic
Onofrite	$(HgS)_{20-80}(HgSe)_{80-20}$	Cubic
Goldamalgam	Au_2Hg_3	Fine-grained aggregates
Potarite	Hg_2Pd_3	Cubic
Schuetteite	$HgSO_4.2HgO$	Trigonal - Trapezohedral
Schwatzite	$(Cu,Fe,Hg)_{12}SbS_{13}$	Cubic
Terlinguaite	Hg_2OCl	Monoclinic
Tiemannite	$(HgS)_{20}(HgSe)_{80}-HgSe$	Cubic

Mercury Distribution in Rocks

The principal mercury occurrences in rocks are related to the Tertiary-Quaternary volcano-orogenetic periods that affected the Pacific, Himalaya and Mediterranean, forming epigenetic and/or syngenetic deposits.

Mercury is commonly associated to epithermal Zn-Pb-Cu Precambrian deposits, which contain the element in low concentrations. The *Kupferschiefer* and Mississippi Valley are typical examples of these deposits (Mitra, 1986).

Saukov (1946) has demonstrated that the mercury deposits are genetically associated with the decreasing intensity of the magmatic conditions during the igneous rocks formation. Saukov (1946) reported Hg concentrations values ranging from $n \times 10^{-4}$ to n % (n varies between 1 and 9) for magmatic rocks, pegmatitic rocks, pneumatolic deposits, epithermal deposits and mesothermal deposits. Mason (1966) reported the following Hg content values: igneous rocks – 80 ppb; schists - 400 ppb; calcareous sandstones - 30 ppb. McNeal and Rose (1974) determined the Hg concentration in the following rocks from Pennsylvania, USA: sandstones (11 samples) – 6.2 ppb; schists – 9 ppb; limestones – 9 ppb (Andersson, 1979).

The variability of the mercury content is several rock types is shown in table 3, which indicates that the schists exhibit the highest values, mainly when are associated to organic matter, as the bituminous schist (Wedepohl, 1979).

Vinogradov (1962) has reported an average Hg concentration of 4×10^{-5} % in schists and of 8×10^{-6} % in igneous rocks. The Hg enrichment in schists was attributed to *phytoplankton* that contains the presence of Hg-adsorber organisms. Vinogradov (1962) also considered that the constituents in hydrothermal processes (Cl, B, F, As and NH_3) are extensively leached to a great variety of rocks, inclusive the schists. Dana (1976) recognized the presence of mercury in magmatic deposits of hydrothermal origin that are accompanied or not by residual solutions providing from the fractionated crystallization of a low temperature magma; during the process, it would be possible occur impregnations and/or filling of veins close to recent volcanic rocks or in thermal springs and geysers too.

Mercury Mobilization

In the several geological environments, the oxidation process is the key factor on the concentration of chalcophile elements, inclusive mercury. This element may occur as Hg^o, Hg_2^{2+} and Hg^{2+}, forming several compounds, according to the physical-chemical conditions of the environment (IPCS, 1991).

The oxi-reduction potential of the mercury valences is within the normal range of the geological environments. The metallic mercury (Hg^o) exhibits low solubility in water (about 100 ppb) (CETEM, 1991a). It is the dominant specie in aquatic environments, being produced by the reaction: $Hg^{2+} + 2\,e^- = Hg^o$ (aq)

The oxidation reactions involving the metallic Hg to their simpler ionic species Hg^{2+} and Hg_2^{2+} have standard Gibbs free energy of formation corresponding to + 39.3 and + 36.5 kcal/mol, respectively, and, therefore, they are not thermodynamically spontaneous. However, these data are related to chemically pure environments, certainly occurring modifications when natural environments are taken into account, as there are actions of other dissolved ions, climatic changes, bacterial actions, and presence of organic matter.

The oxidation of metallic mercury according to the reaction $Hg_2^{2+} = Hg^o$ (aq) + Hg^{2+} may be a consequence of the following factors (CETEM, 1991a): adsorption of clay minerals, aqueous iron and manganese oxides, and organic matter; precipitation like oxides, hydroxides or sulfides; adsorption by biota; reaction with methylcobalamin formed by bacteria, resulting from methylmercury ions; formation of methylmercury complex ions with chloride, carbonate, sulfate and hydroxides; complexation with others organic groups.

Table 3. The mercury content in some types of rocks. According to Wedepohl (1979)

ROCK	Hg (ppb)
Granite	3.5 (1.4 - 281)
Diorite	5.0 (0.4 – 19.7)
Diabase	3.9 (0.2 – 17.7)
Sandstone	2.5 (0.8 – 6.0)
Schist	5.9 (0.9 – 33.5)
Bituminous schist	234 (31.9 - 340)
Limestone	6.0 (0.8 – 31.2)
Gneisses	2.6 (1.5 – 3.8)

MERCURY ADSORPTION

The adsorption processes in clay minerals present in soils and sediments occur at accentuated reaction velocity, mainly when are involved kaolinite, montmorillonite, illite and chlorite. These clay minerals have been extensively investigated over the last decades because of the presence of stable secondary minerals formed by the decomposition of other aluminosilicates. The recently formed minerals constituents of the clay minerals are classified as phyllosilicates, have structures in layers and organic composition of high specific surface area. They are hydratated aluminum silicates, whose composition facilitates the substitution and ionic exchange in the crystal lattice by the adsorption of heavy metals ions (Mason, 1971; Krauskopf, 1972; Frenet, 1979; CETEM, 1991a; Summerfield, 1994; Krauskopf and Bird, 1995). The ionic exchange is characterized by the balance of the superficial charge due to the bond break of the aluminosilicates during the hydrolysis and hydration processes, which also can occur when the ions are not completely coordinated (CETEM, 1991b).

The adsorption of metallic mercury by organic matter depends on the physical-chemical characteristics of the sediments, existing a good relationship among the values of the energy constant for the metal-sediments bond, organic matter content and grain size of the (Ramamoorthy and Rust, 1976, 1978; Moore and Ramamoorthy, 1984). The concentration of the total mercury in sediments increase with the decreasing of particle size; values of 70% and above have been pointed as representative of its adsorption by the organic matter (Loring, 1975; Silva, 1996). Therefore, clay minerals, iron oxides and hydroxides, silicic acid and organic matter occurring in sediments are compounds with high specific surface area that are able to adsorb cations from solutions by ion exchange. The ion exchange process is based on the sorbent properties of anionic sites (negatively charged) relatively to cationic ones (positively charged) that are exhibited by SiOH, $AlOH_2$ and AlOH groups (in clay minerals), FeOOH groups (in iron oxides and hydroxides), HS sulphydryl groups, COH carboxylic and phenolic groups (in organic matter).

The organic matter has several ways of interaction with metallic ions in the aquatic environment and this implies that the heavy metals affinity by organic substances is significantly important in the biogeochemical cycle. The humic acids are among the decomposition products of the organic substances, and they have accentuated activity of complexation, chelation and adsorption of metallic cations when dissolved in water, becoming strong reducing agents of metallic cations.

Chapter 3

The Madeira River Basin at Rondônia State, Brazil

The local and international specialized literature about the Amazonian area has increased over the last decades due to its importance in the global scenario. This chapter describes some general features of the area investigated that are relevant to understand the monitoring program held for surface waters, bottom sediments, and fishes.

Location of the Investigated Area

The area focused for investigating the mercury behavior in the environment is situated at the Porto Velho region and neighborhood, Rondônia State, Brazil (figure 1). It includes the sector between Teotônio and Santo Antônio waterfalls at the Madeira River that is in the west-southwest direction from Porto Velho city, at the norhwest portion of Rondônia State (figure 1). The approximate surface area of this region is 260 km^2, including part of the urban perimeter of the Rondônia State capital (Porto Velho city) and being delimitated by the following coordinates: North- 63°51' 34"/ 8°48'16"; South- 63°54'49"/8°43'31"; East- 64°5'24"/8°51'00" and West- 64° 2'10"/8°55'37".

The main terrestrial routes to access the area are: BR-364 road (Porto Velho - Rio Branco direction), BR-319 road (Porto Velho – Manaus direction) and Madeira-Mamoré railway (sector between Porto Velho and Santo Antônio village). The access to the area by fluvial transport occurs by medium and large-sized carriers until Santo Antônio fall and small-sized carriers until Teotônio fall.

Santo Antônio fall is about 6 km distant from Porto Velho city if the route is realized by Madeira-Mamoré railway, however, the distance is 31 km if the path is performed by roads (18 km by BR-364 paved way and 13 km by unpaved secondary ways). Santo Antônio fall only may be reached by foot on using BR-319 road (Porto Velho – Manaus direction).

SYNTHESIS OF THE REGIONAL GEOLOGY EVOLUTION

The area investigated is located at the southwestern portion of the Amazonian craton (Almeida, 1978, 1984) and within the limits of the Madeira sub-province (Amaral, 1984). The compartmentation and geotectonic evolution of the Amazonian craton have been evaluated by models of different conceptions. The first utilizes a fixist approach, where the Amazonian craton represents an enormous continental mass existing since the Archean/Lower Proterozoic and that was successively re-worked and re-activated during the Intermediary and Upper Proterozoic (Almeida, 1974, 1978; Amaral, 1974, 1984; Santos, 1984; Hasui et al., 1984; Hasui and Almeida, 1985; Costa and Hasui, 1991). The second model utilizes a mobilist conception, where the craton evolution is characterized by episodic crustal accretions during the Proterozoic and is associated to the geodynamic development of the region that acted in successive periods of three mobile belts (Cordani et al., 1979; Gibbs and Barron, 1983; Gaudette and Olszewski, 1985; Teixeira et al., 1989; Litherland et al., 1985, 1989).

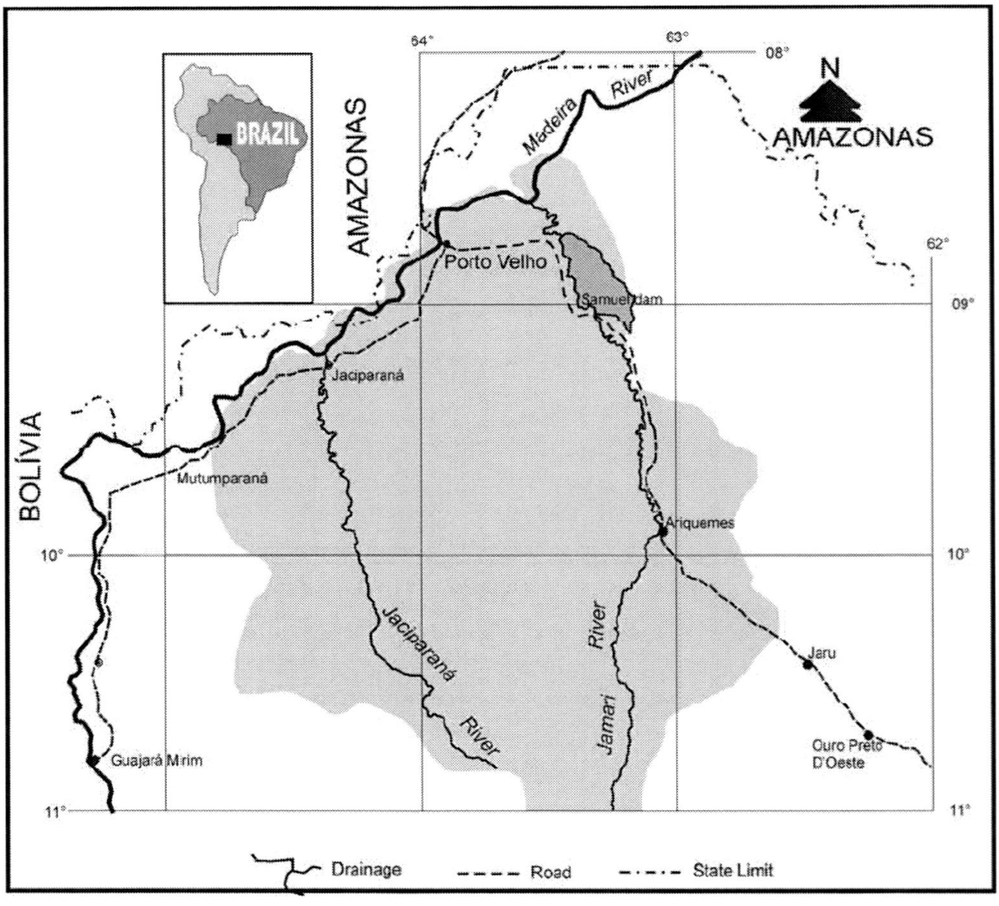

Figure 1. Location of the area for sampling surface water, sediments and fishes in Rondônia State, Brazil. According to Gali (2002).

The fixist model considers that the last important orogenetic event of the Amazonian craton occurred during the Trans-Amazonian cycle (~ 2,000 Ma) when the region was submitted to three important re-activations within the time interval of 1,900-1,000 Ma, i.e.:

a Uatumã, from 1,900 to 1,700 Ma (Santos, 1984; Santos and Loguércio, 1984) or Paraense, from 1,700 to 1,550 Ma (Amaral, 1974);
b Parguazense, ~ 1,500 Ma (Bellízia, 1972) or Madeirense, from 1,400 to 1,250 Ma (Amaral, 1974);
c Rondoniense, from 1,050 to 900 Ma (Amaral, 1974).

These re-activations were accompanied by intense magmatic processes that produced accentuated faulting, volcanism and plutonism with posterior formation of sedimentary covers and detritic deposits (Shcheglov, 1970; Amaral, 1974).

The other model (Cordani et al., 1979; Cordani, 1981; Teixeira et al., 1989) considers that the geodynamic development of the continental crust results from the action of four active and distinct mobile belts during successive periods, which are roughly oriented according to the NW-SE direction, characterizing a chronostructural province named Central Amazon. This province is constituted by granite-greenstone terrains with undeformed volcano-sedimentary covers and anorogenic intrusives whose ages are between 1,900 and 1,500 Ma. The Maroni-Itacaiúnas Province (2,250 – 1,900 Ma) includes granite-greenstone terrains from the Lower Proterozoic and granulitic-gneissic terrains (Santos et al., 1987; Gibbs and Barron, 1983). The south margin of the Central Amazon province is related to three different tectonic provinces:

a Rio Negro-Juruena province (1,750 – 1,500 Ma), constituted by granitoid and gneissic rocks (Tassinari, 1981);
b Rondoniana province (1,450 – 1,250 Ma) (Teixeira and Tassinari, 1984);
c Sunsás province (1,100 - 900 Ma) (Litherland et al., 1989).

Geological Description of Rondonia State and Neighborhood

The geological map of Rondônia State and vicinity is shown in figure 2. The Jamari complex is the oldest unit of the Amazonian craton, according to the stratigraphy. It is overlaid in polymetamorphic terrains constituted by gneissic rocks tending to intermediary and high methamorfic grade, granitoids and, less frequently, paragneisses, migmatites, amphibolites, metabasites and granulites (Isotta et al., 1978).

Because of the sparse geochronological data, there is controversy on several ages for the rock formation (1,800 – 1,500 Ma), as many propositions are speculative and coupled to two models for the evolution and geotectonic compartmentation of the Amazonian craton. The Jamari complex (figure 2) is constituted by: ortho and para-derived high grade gneisses, tectonically *imbricated; foliated* granitoids, locally characterized as orthogneisses (falls in Samuel hydropower) that include granitic orthogneisses (secondarily granodioritic), with variable deformation generated in a compressive tectonic environment caused by ductile shearing of low and intermediary metamorphic grade.

The Paleoproterozoic has been subdivided in two supracrustal sequences by Scandolara et al. (1996), i.e. the Nova Brasilândia metavolcano-sedimentary sequence (Comemoração metamorfites) and the Roosevelt metavolcano-sedimentary sequence. The first sequence is composed by psamopelitic and chemically exalative rocks, ocean- floor metabasalts, and pyroclastics rocks that have deposited in a subaqueous environment as a consequence of low grade conditions. The second sequence is constituted by a metamorphized lithological group under low grade conditions, with large dominance of intermediary/acid pyroclastics lavas and subordinated intercallations of clastic sediments, dominantly pelitic.

According to Scandolara et al. (1996), the Mesoproterozoic granitoids (1,660 – 1,300 Ma) are represented by chemically subalkaline intraplate granites, which have been grouped together according to the geochronology and dominantly rapakivic textural-compositional characteristics in the following types: granites, charnockites, mangerites and associated basics (Serra da Providência suite); subalkaline-alkaline granitoids (Teotônio/Santo Antônio massif intrusive suite); young rapakivi granites (Bettencourt and Dall'Agnol, 1987) formed from a polyfase granitic pluton; associated basics (São Lourenço/Caripunas complex, Siriquiqui basics).

Souza and Marques (1974), Lima et al. (1978) and Isotta et al. (1978) have associated the Santo Antônio granites with those of the Massangana facies of the Rondônia intrusive suite. The Serra da Providência granite with rapakivi texture has been described by Adamy and Romanini (1990).

One occurrence of alkaline rocks in Rondônia State at Teotônio fall, Madeira River that had been previously identified by Evans (1906) and Verschure and Bon (1972) was considered as a specific unity by Leal et al. (1978). This occurrence reflects a 4-km diameter circular intrusion and represents an alkaline emanation coupled to an autonomous activation process of the platform that is characterized by an association of syenite, egerine-monzogranite and egerine-granite. Adamy and Romanini (1990) named the Morrinhos complex as the group of rocks between the São Patrício island and Morrinhos fall, whereas Payolla (1994) and Scandolara et al. (1996) designated the Teotônio intrusive suite as the area of occurrence of these rocks.

The Palmeiral formation represents a Meso/Neoproterozoic (1,300 – 1,220 Ma) sedimentary cover of platformal environment that is constituted by conglomerates, arkosian sandstones and pelites outcropping in vast fault-conditioned areas derived from fluvial systems with marine influence.

The Neoproterozoic in the area is constituted by: alkaline basalts (subordinately diabasic) from Nova Floresta formation, which are related to distentional processes that originated the Meso/Neoproterozoic basins; young Rondonian granites corresponding to anorogenic rocks represented by alkaline feldspar granites, sienogranites and monzonites, which were grouped together as a consequence of their alkaline chemical characteristics and that are strongly conditioned to a ruptile recurrent tectonics.

The Pimenta Bueno formation is a Paleozoic sedimentary cover represented by an interdigited association of marine lithofacies, continental facies, glaciocontinental facies and glaciomarine facies that are constituted by shales, arkosian sandstones, diamictites, pelites with variable types of clasts and limestones.

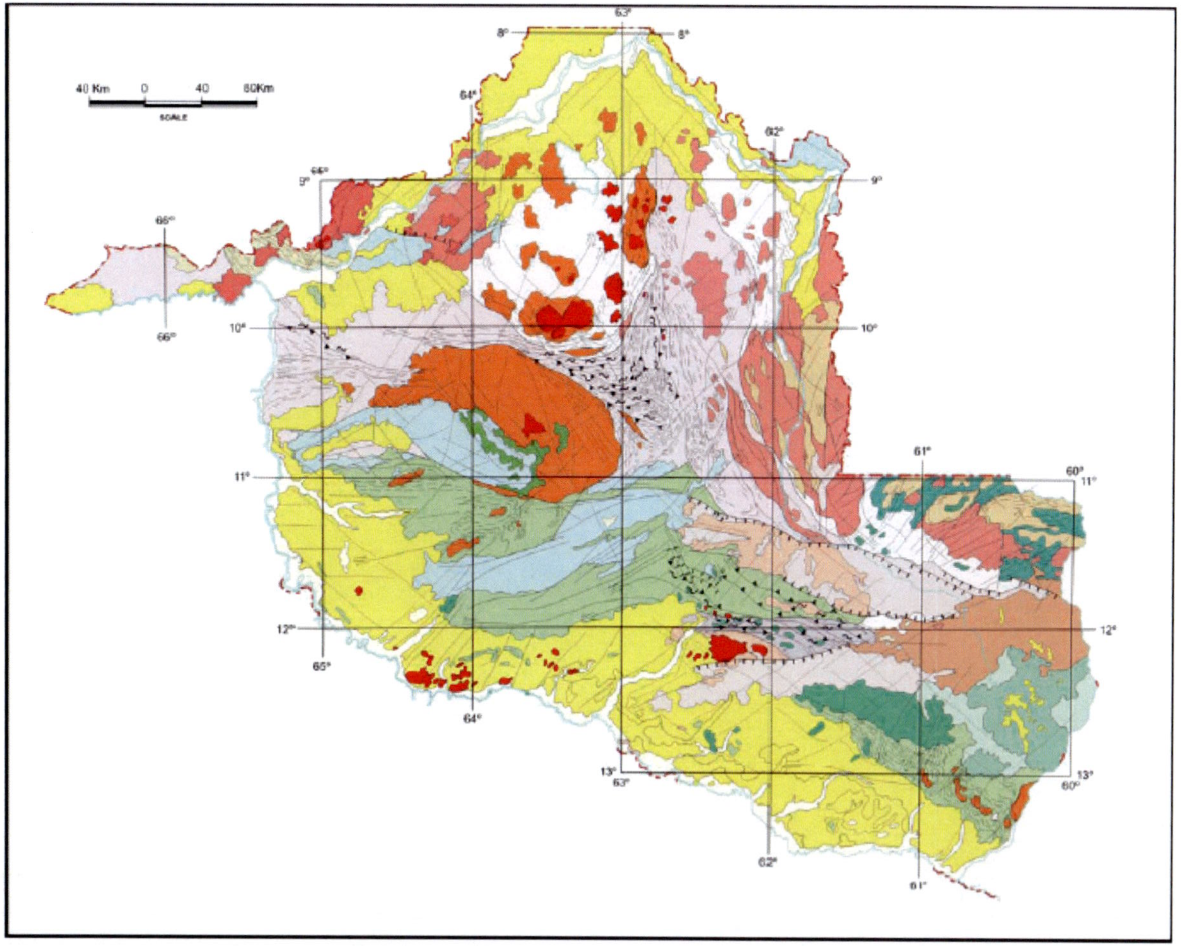

Figure 2. Continued on next page.

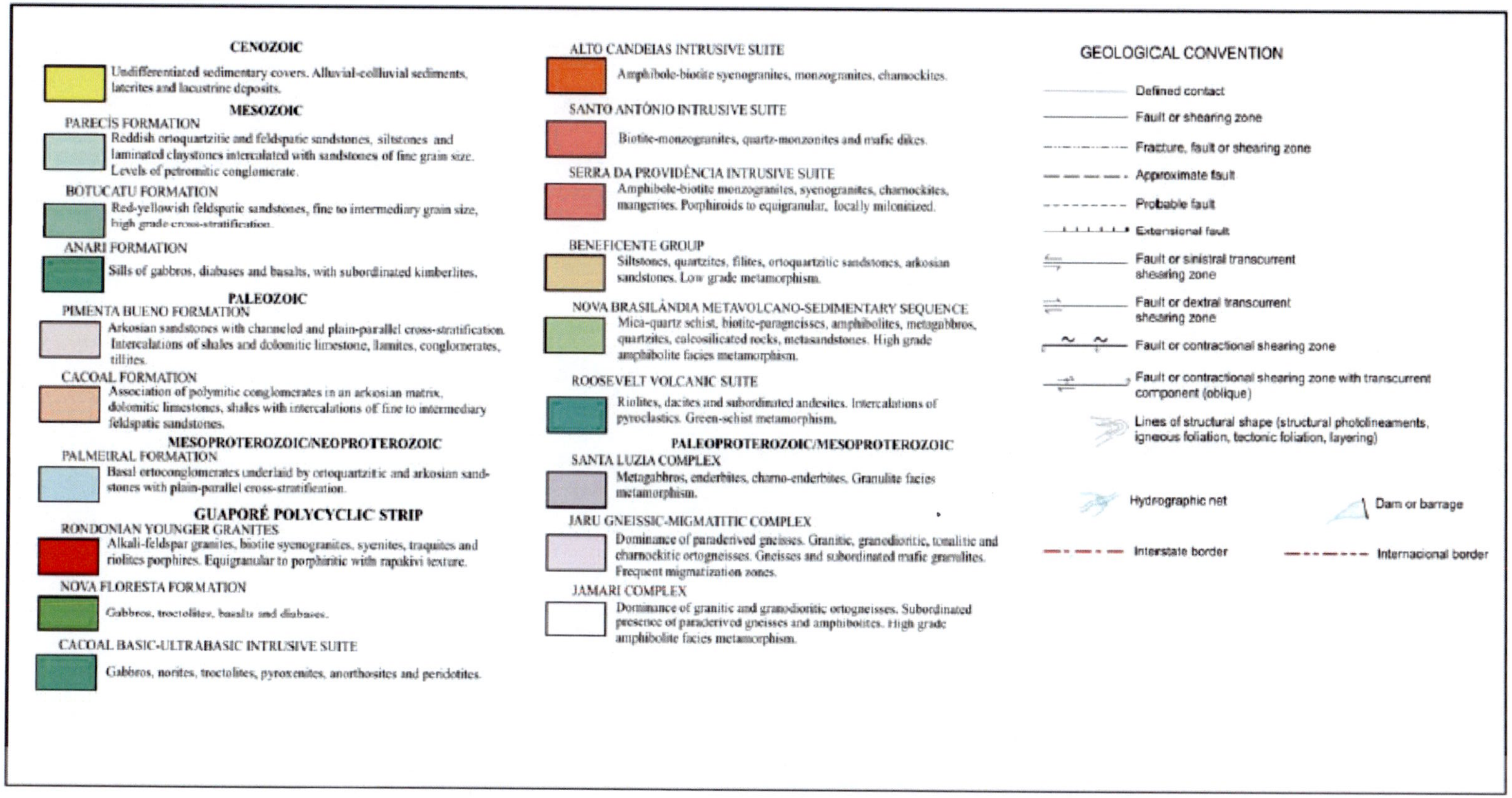

Figure 2. The geological map of Rondônia State and nearby. According to Scandolara et al. (1996).

The Mesozoic is represented in the area by the Anari and Parecis formations (Scandolara et al., 1996): the first corresponds to a unit of basic and ultrabasic rocks, diabases and kimberlites related to an environment of continental distension. The second comprises a sedimentary cover constituted by thick packages of whitish and reddish sandstones into which are subordinated conglomerates and pelites related to a desertic system, with facies characteristics of fluvial, eolian and fluvio-eolian sub-environments.

The Cenozoic sedimentary cover is named Jaciparaná formation in the region between Porto Velho and Abunã and developed mainly in the plains (areas of the large rivers) and plateaus (basement areas), where are concentrated lateritic terrains, clayey-sandy sediments deposited in the Upper Pleistocene and unconsolidated alluvial deposits (Payolla, 1994).

Additional Geological Informations

The geological work of Evans (1906) was pioneer on the study of the Madeira River region and contains descriptions of the rocks exposed in the waterfalls of Madeira, Mamoré and Beni rivers, i.e. syenitic rocks cut by coarse granites and aegirine-augite-olivine basalt veins in Teotônio fall, and coarse hornblende-biotite granites in Santo Antônio fall.

The studies held by Lobato et al. (1966) characterized the region considering tectonostructural, stratigraphic and lithological aspects, as well of the major mineral occurrences, mainly the cassiterite deposits. Pegmatoid pinkish granitic rocks have been identified along the Madeira-Mamoré railway tracks, close to Santo Antônio fall.

Verschure and Bon (1972) stratigraphically characterized the regional granites as anorogenic vulcano-plutonic complexes and determined a Rb-Sr age of 1,305 Ma for a coarse biotite granite in Santo Antônio fall. They also identified microfractures filled by aegirine-albite, magnesioriebeckite and stilpnomelane cutting granitic rocks in Teotônio fall. Souza and Marques (1974) verified that the granites of Santo Antônio fall constitute a batholith whose outcrop area corresponds to about 110 km along the Madeira River, since Porto Velho up to Caldeirão fall. They have been referred to Santo Antônio porphiroblastics and consist on pinkish hornblende-biotite granites whose granulation is coarse and texture is rapakivi-dominated.

Leal et al. (1978) provided the first integrated geological map of Rondônia State in their report realized to RADAM project. They considered that the granitic rocks of Santo Antônio fall belong to Xingu complex, whereas the granites, syenites and monzonites of Teotônio fall were named Teotônio alkalines.

Leal et al. (1978) have included the Teotônio alkalines in the Xingu complex based on the descriptions performed by Evans (1906). Verschure and Bon (1972) took into account the relationships identified between the field evidences and the results obtained of petrographic and geochronological analyses (Rb-Sr ages of 1201 ± 104 Ma and 1249 ± 60 Ma) in order to justify the term of Teotônio alkalines to the 4-km diameter circular syenitic intrusion. However, Bettencourt et al. (1995) redefined the Teotônio intrusive suite based on petrographic, geochemical and geochronological analyses, as they determined Rb/Sr and U/Pb ages ranging from 1,406 to 1,370 Ma.

Isotta et al. (1978) subdivided the granitoids of the region in three major groups, i.e. associated to the crystalline basement, genetically related to rapakivi, and Rondonian granites. The term Rondônia intrusive suite was applied to the granitic rocks, whereas the granites of Santo Antônio region were individualized and referred to Porto Velho massif, integrating the Rondônia intrusive suite.

ELETRONORTE (1987) has displayed a regional geologic map and semi-detailed maps from Santo Antônio and Teotônio falls. It was identified the presence of ortogneisses and paragneisses megaxenolithes of the crystalline basement in the coarse granitic rocks of Santo Antônio fall, as well the presence of syenitic and monzonitic rocks in the Madeira River channel between São Patrício island and Teotônio fall, indicating that the exposed area of the Teotônio alkalines is much greater than that proposed by Leal et al. (1978).

Adamy and Romanini (1990) organized the explanation text of the Brazilian program for basic geological mapping, where there is reference to the geological mapping of the Porto Velho, Mutumparaná, Jaciparaná and Abunã sheets, inclusive the definition of the Morrinhos complex that corresponds to syenites, monzonites and granites which are correlated to the Teotônio alkalines based on petrographic and petrochemical analyses. The granitic body of Santo Antônio fall was designated Santo Antônio massif and correlated with the Serra da Providência granite due to their rapakivi texture.

Payolla (1994) presented the results of the geological, petrographic and geochemical characterization of granitoids and syenoids from Teotônio (Teotônio intrusive suite) and Santo Antônio falls. The coarse-intermediary grained granites occurring at the Porto Velho-JaciParaná region and that are porfiroids/equigranular in character were designated Santo Antônio-type granites (Payolla, 1994).

Figure 3 shows the geological map of the area between Teotônio and Santo Antônio falls in Madeira River, which was elaborated by Payolla (1994) according to petrographic and geochemical analyses of rocks and further modified by Silveira (1998). The following lithologies occur at the Porto Velho city region and nearby:

- a Recent alluvial sediments (Qh) – they are related to holocenic alluvium that are conditioned to the Madeira River channel. They are formed by alluvial deposits constituted by pebbles, fine sand, silt, unconsolidated clays and organic matter that morphologically comprise the fluvial terraces between Teotônio e Santo Antônio falls.
- b Jaciparaná Formation (Qph) – formed by vast areas of Cenozoic cover, discordantly disposed over rocks of the Jamari complex, which are primarily constituted by clayey and silty-clayey sediments and secondarily by sandy sediments of variable grain size.
- c Laterites (TI) – mainly formed by two dominant types: a) immature or little evolved laterites, constituted by soil horizons (spherulitic and mottled concretions), concretionary horizon, mottled horizon and pale horizon; b) mature or evolved laterites, represented by bauxitic laterites covered by colluvium and pleistocenic sediments that are related to the main drainage of the region.
- d Santo Antônio massif (PM γ_{SA}) – constituted by numerous facies of granitoids that are secondarily intercalated by diabase dikes and distributed as follows: association of hybrid rocks (porphiritic facies I, porphiritic facies II and breccia); association of intermediary granitoids (medium pink and gray facies); association of fine gray granites (fine gray facies) and association of coarse granites (pink I and pink II facies).
- e Teotônio intrusive suite (PM γ_{TE}) – constituted by numerous facies of granitoids and syenoids that are subordinated to dikes of mafic rocks. The major facies are identified as follows: coarse pink association (coarse pink facies I, II and III);

intermediary banded association (intermediary equigranular dark pink banded facies I) and coarse massif association (equigranular and porphiritic coarse massif laminated facies I and II).

f Jamari complex (PM_j) – basically formed by polimetamorphic rocks like gneisses, quartz-diorite ortogneisses, calcosilicated rocks and syenogranites, existing granitic and/or granodioritic zones affected after the Trans-Amazonian cycle.

CLIMATE

The Rondônia State exhibits equatorial climate controlled by a zonal microclimate and a system of low latitude disturbed currents, with heavy rains and accentuated relative air humidity during the whole year. The regional climate, according to the Köppen classification, is Am-monsoonal tropical, transitioning to Aw-tropical rainy climate. The region has two very distinct climatic periods (Nunes, 1999): a) the wet season (average rainfall = 250 mm/month), occurring between October and April; b) the dry season (average rainfall = 20 mm/month), occurring from May to September and characterized by high thermal amplitude and reduced clouds formation (Nunes, 1999). The climate is hot under the perspective of the thermal regime; the average annual air temperature is 25.5 °C, whereas the minimum average annual value corresponds to 20.7 °C, and the maximum to 32.2 °C. The relative air humidity ranges from 74.2 to 90.8%. The minimum, maximum and average annual precipitation corresponds to 1250 mm, 2000 and 1625 mm, respectively (IBGE, 1993).

VEGETATION

There is an accentuated diversity of vegetal species in Rondônia State, which are represented by: opened ombrophilous forest (transition forest), savannas and areas of ecological tension and anthropic actions.

The opened ombrophilous forest is characterized by well-spaced arboreous specimens, with several tall trees (25-50 m high) that are easily identified because of the falling leaves during the dry season and of the richness in lianas and epiphytes. The median arboreous stratum is characterized by leguminous plants, Moraceae, Sapotaceae and Lecythidaceae, with the presence in some areas of *Orbignya phalerata* (babasu palms), *Oenocarpus bataua* (Patauás) and *Maximiliana regia* (Ingás). The lower arboreous stratum is formed by small trees and young specimens.

The cerrado fields and savannas occur predominantly in southern and southwestern sectors of Rondônia State, in vast areas covering the major part of Serra dos Pacáas Novos and Chapada dos Parecis. The cerrado fields occupy areas of plain terrains covered by clayey Cenozoic sediments, as well areas of accentuated relief sculpting Precambrian rocks. In general terms, the vegetation of the cerrado fields is typically characterized by the presence of short trees that are isolated or grouped together on a graminea grasses cover. The woody plants exhibit well-protected bud leaves, thick cork layers and well developed leaves with stomates protected by trichomes. The most common specimens are *Qualea grandiflora* and *Qualea parviflora* (paus-terra), *Salvertia convallariodora* (colher-de-vaqueiro), *Dimorphandra mollis* (faveiro), *Stryphnodendron barbatiman*

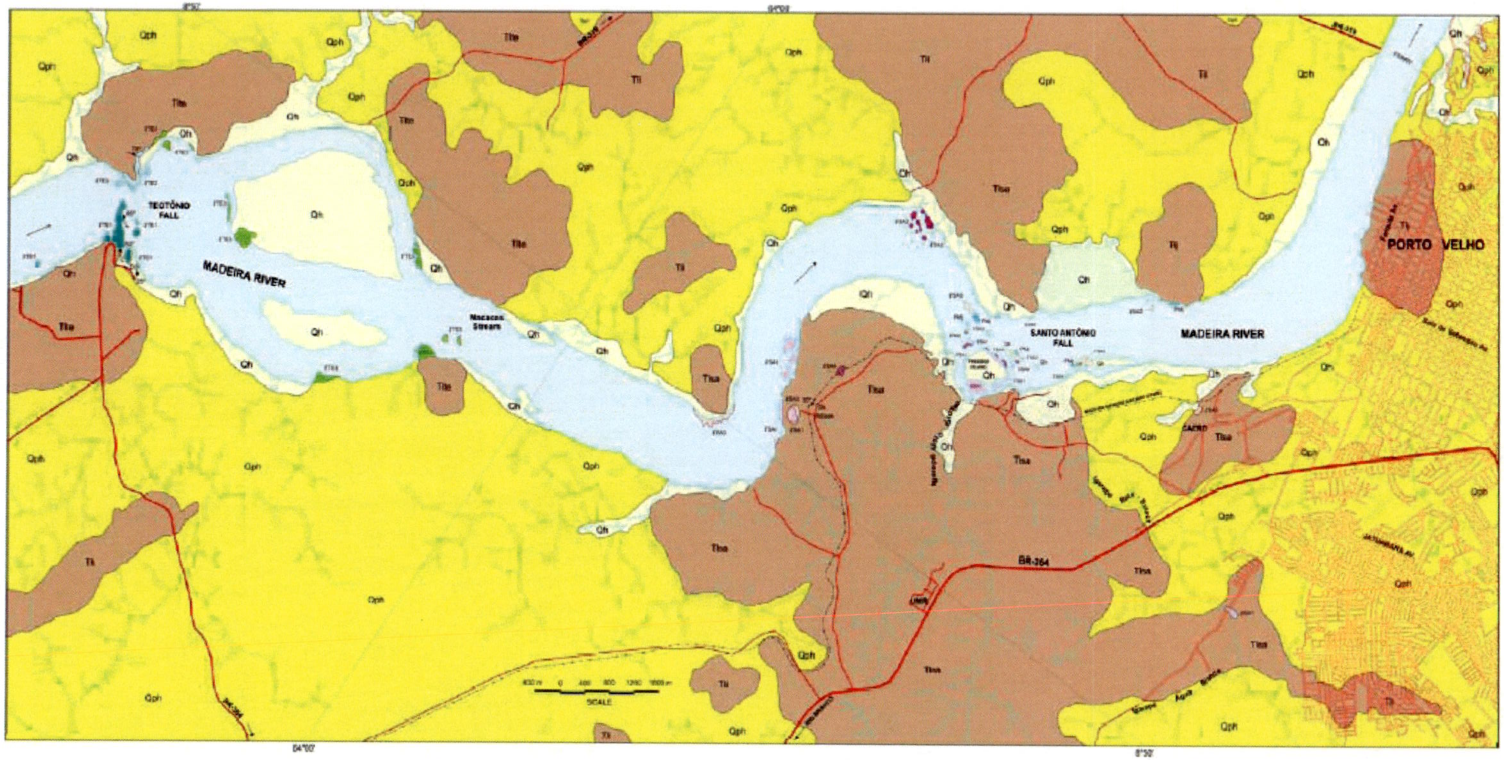

Figure 3. Continued on next page.

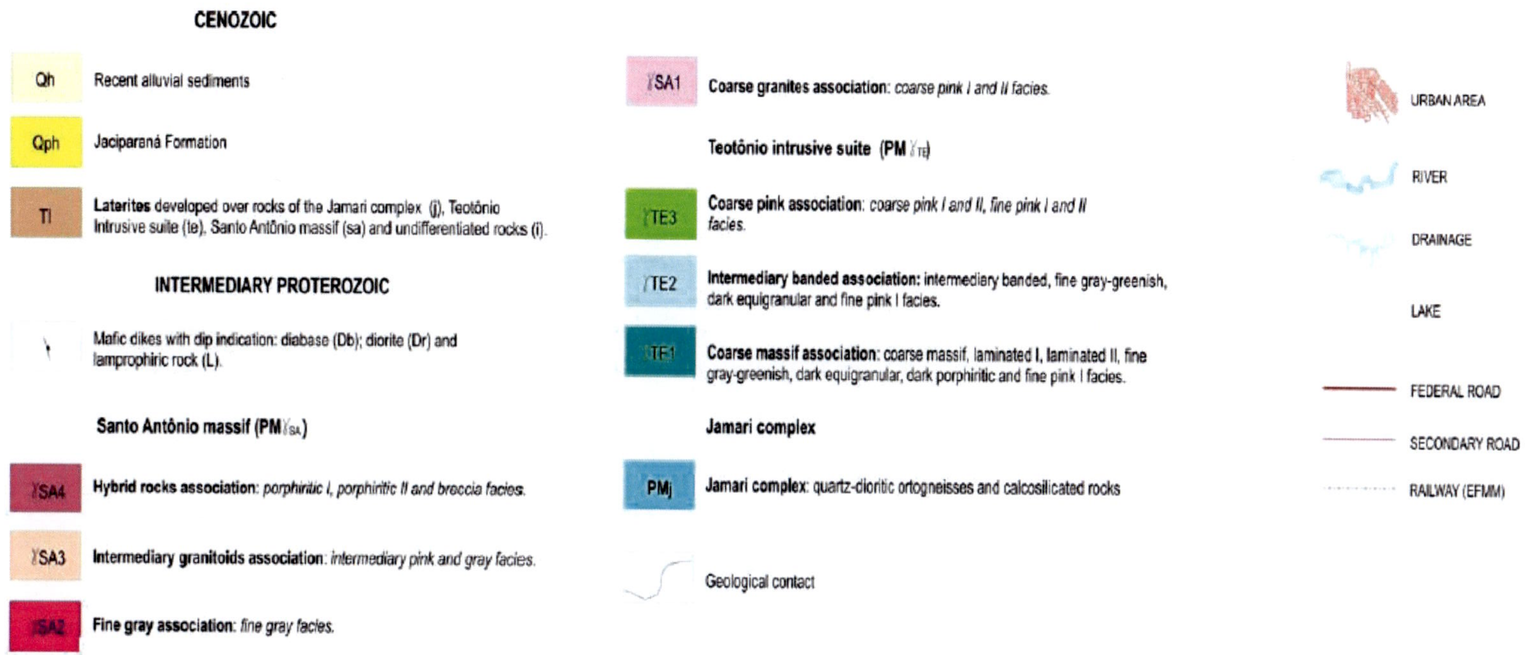

Figure 3. The geological map of the area comprising Teotônio and Santo Antônio falls at Madeira River, Porto Velho, Rondônia State, and nearby. According to Payolla (1994) and Silveira (1998).

(barbatimão), *Curatella americana* (lixeira), *Caryocar brasiliensis* (pequi), *Magonia pubescens* (timbó) and *Piptadenia macrocarpa* (angico-preto), among others (IBGE, 1993; DNPM, 1978).

The areas of ecological tension and anthropic actions are located in the left side of Madeira River, corresponding to contact and transition zones from cerrado to forest. The areas of anthropic actions correspond to regions established for the development of pasture and agriculture that were occasionally abandoned and succeeded by secondary vegetation, originating *capoeirões*.

HYDROGRAPHY

The hydrography in Rondônia State is controlled by Madeira River. In terms of geomorphological aspects, the hydrographic system of Madeira River comprises the following areas: the Central Brazilian Plateau in the right side, the Guaporé/ Mamoré valley and the Beni River at the Bolivian Cordillera Occidental, Cochabamba Mountain, in the left side.

The waters of Madeira River are dominantly clear, except during the rainy periods when occurs increase in turbidity due to the raising of the suspended solids load, as the Secchi transparency may reach values between 0.41 and 2.32 m (Ribeiro and Santos, 1983). The total drainage area is approximately 1.3×10^6 km^2 and the annual discharge corresponds to 1.4×10^{12} m^3/year (Goulding, 1979). The channel of Madeira River cuts Precambrian rocks in Rondônia State and the difference in altitude between 3 and 9 meters allows the formation of 21 waterfalls along its course. However, the waters also drain Cenozoic sediments, where there is no presence of falls but mainly floodplain areas that favor the occurrence of flooding during the raising of the water level. The water level height of the Madeira River is directly linked to the rain regime, with the highest values occurring in the wet season (between October and April) as illustrated in figure 4.

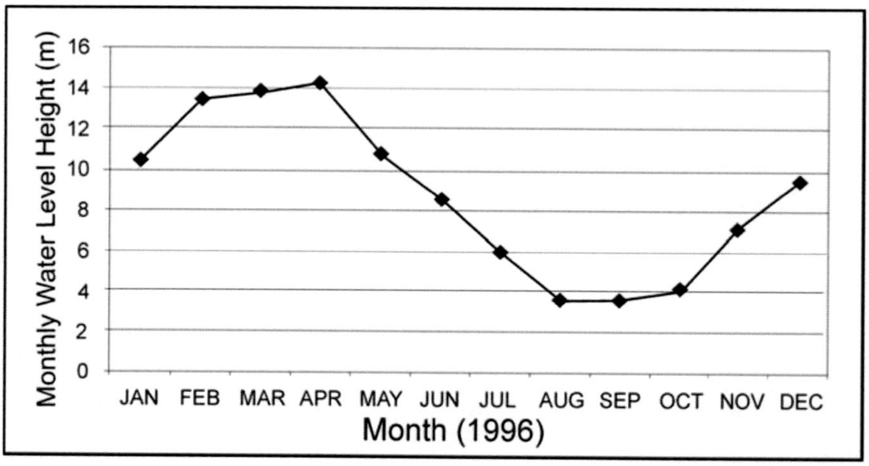

Figure 4. The average monthly water level height of the Madeira River during 1996. According to Nunes (1999).

Geomorphology

The Rondônia State is mainly characterized by two well-defined morphostructural units: the Amazon South Dissected Plateau, identified by Barbosa and Pinto (1973), and the Amazon Occidental Lowered Plateau (Isotta et al., 1978; Melo et al., 1978). The Amazon South Dissected Plateau is composed by Precambrian sediments and granitic rocks, whose major features are: smooth relief, elongated, steep scarps and slightly tabular top with mean height of 250 meters. The border of the tabular relief forms narrow V-shaped valley frequently marked by deep moulding. The Amazon Occidental Lowered Plateau covers vast areas in advanced stages of planation that are constituted by granites, gneisses and amphibolites, dissected relief covered by little consolidated sediments and laterites. This assemblage is characterized by an eroded relief, with some residual elevations of plained top, originating a tabular interfluvium that may reach 175 meters height.

Chapter 4

MEASURING MERCURY

The atomic absorption spectrometry is a successful technique utilized for quantifying mercury. This chapter describes the main steps conducted for establishing the methodology at the Environmental Geochemistry Laboratory belonging to UNIR (*Fundação Universidade Federal de Rondônia*), Porto Velho, Rondônia State. The method was applied on the evaluation of the mercury content in sediments and rocks collected between Teotônio and Santo Antônio falls, close to Porto Velho city.

MAJOR FEATURES OF THE INSTRUMENTATION

The atomic absorption spectrophotometer utilized at UNIR is a specific instrument utilized for the Hg quantification, based on the cold vapor generation. It is a Flow Injection Analysis - Mercury Hydride System/FI-MH-AAS (FIMS-400) from Perkin Elmer, containing an automatic sampler (AS-90) and microcomputer with software for operating the system. The following parts comprise the equipment: cylindrical cell with quartz window, a Hg lamp (253.7 nm), a specific detector for absorption on the Hg wavelength, a system of flow injection analysis, and a cold vapor generator system. The parameters of calibration utilized in this equipment consisted on an Ar flow of 50 $mL.min^{-1}$, a 3% HCl flow between 9 and 11 $mL.min^{-1}$, and a 0.2-0.5% $NaBH_4$ + NaOH flow between 5 and 7 $mL.min^{-1}$. The concentrations of the analytical solutions for obtaining the calibration curve of the equipment varied between 2 and 40 $\mu g.L^{-1}$, and were prepared from a 2 $\mu g.mL^{-1}$ stock solution preserved with 5% HNO_3 and 0.01% $K_2Cr_2O_7$. Figures 5a and 5b schematically represent the system.

The atomic absorption spectrophotometer coupled to a hydride generator may be considered one of the most important techniques for determining mercury. It offers better sensitivity and faster reactions, being practical for cleaning the reaction glassware; however, its weakness is the relatively high price, becoming difficult the access for many laboratories (Boaventura, 1995). The methodology was established at UNIR on using 24 samples of sediments (standards), whose total Hg concentration had also been measured at the Laboratory of Radioisotopes Eduardo Penna Franca of the Institute of Biophysics Carlos Chagas Filho of the Federal University of Rio de Janeiro (UFRJ) in Brazil. The equipment utilized at UFRJ is the same described by Malm et al. (1989), i.e. a Varian model AA-1475 coupled to a vapor generator Varian model VGA-76. The operational parameters used in this equipment consisted on a sample flow of 8 mL.min^{-1}, a N_2 flow in the gas-liquid separator of 18 mL.min^{-1}, and a $NaBH_4$ + HCl reagents flow of 1 mL.min^{-1}. The model of the absorption cell is cylindrical, and it contains a quartz window with internal unidirectional flow without heating, due to the better sensitivity.

The same procedures were utilized at both laboratories for digestion and extraction for Hg analysis in the standards, which followed the techniques developed by Malm et al. (1989). About 0.5 g of each standard was used, 1 mL of Milli Q water plus 2 mL of aqua regia were added, and heated to 50°C for 5 min. After cooling, 10 mL of Milli Q water plus 4 mL of 5% potassium permanganate were added and heated to 55°C for 30 min. The excess $KMnO_4$ was neutralized with hydroxylamine hydrochlorate 12%. Each aliquot representing the standard was divided in three sub-aliquots.

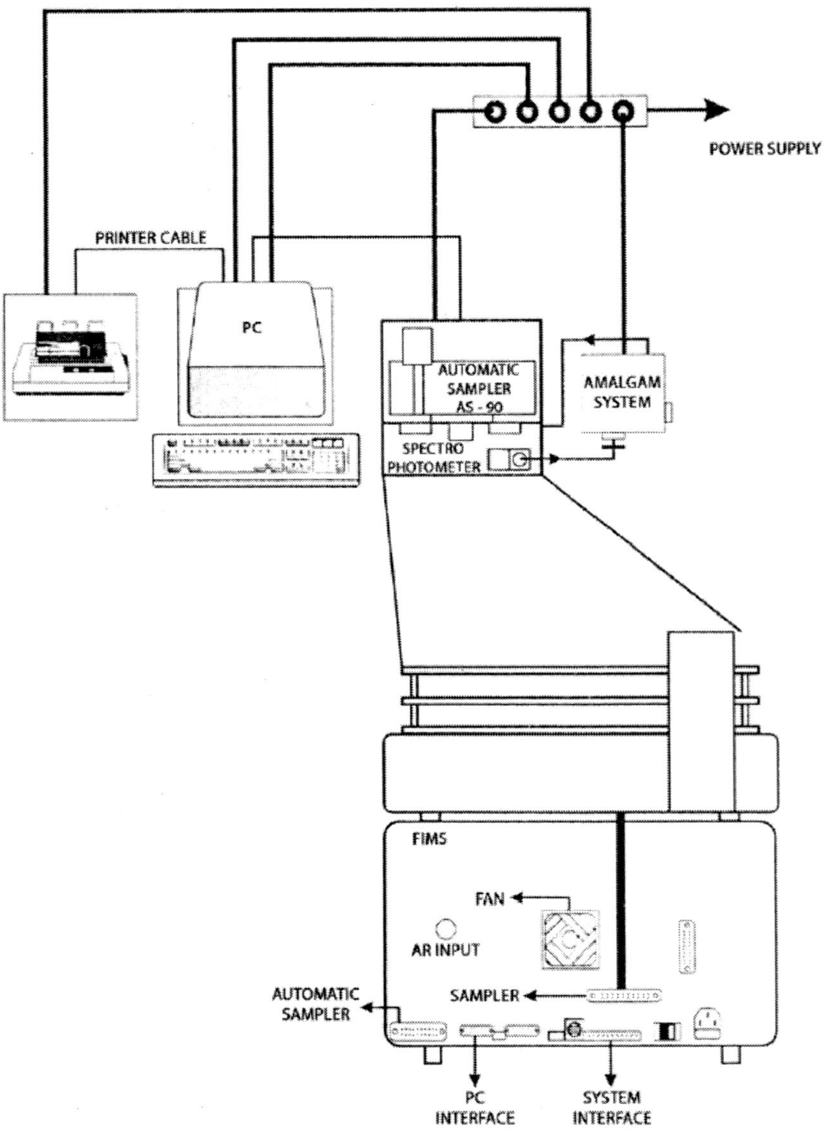

Figure 5a. Simplified block diagram of the FIMS-400 system utilized for quantifying mercury.

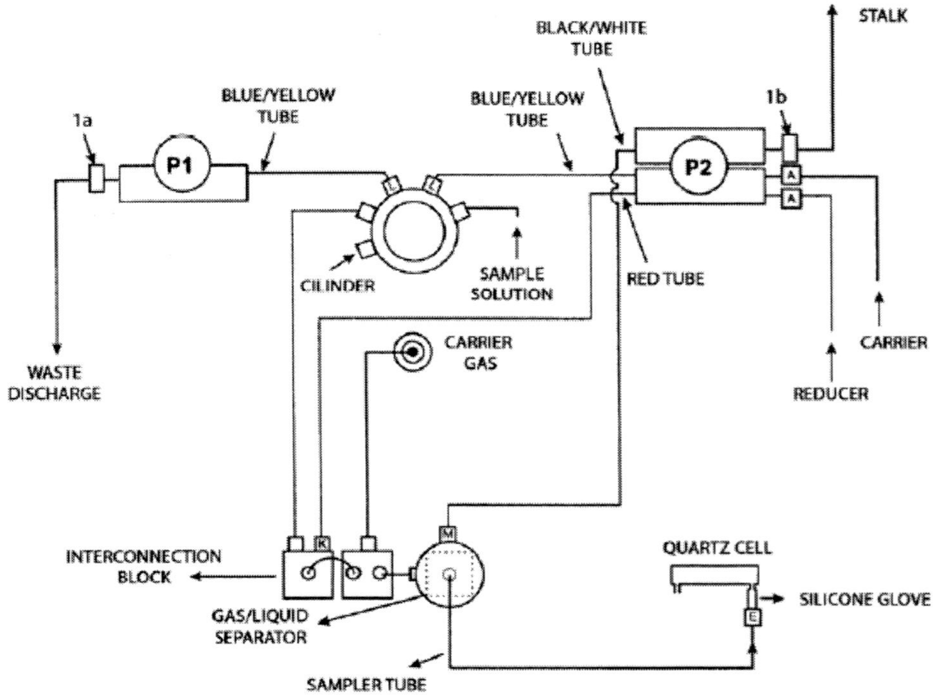

Figure 5b. Simplified block diagram of the flow system in the FIMS-400 atomic absorption spectrophotometer utilized for quantifying mercury.

The 24 samples of sediments consisted on: a) four standards whose Hg concentration had been previously measured by Malm (1991) and intercalibrated at the Food Laboratory from National Food Administration/Holland and Institute for Soil Fertility/Holland; b) one reference standard (sediment pound from National Institute for Environmental Studies/NIES - Japan Environmental Agency); c) nineteen samples with unknown Hg content, whose quantification was performed by the two atomic absorption spectrophotometers.

Calibration of the FIMS System from UNIR

The solutions and reagents were prepared on weighing 120 g of NaCl, adding 500 mL of Milli-Q deionized water into a 1000 mL volumetric flask, and shaking the mixture by about 3 hours. Then, it was added 120 g of $HONH_3Cl$ (hydroxylamine hydrochlorate) to the NaCl solution, the mixture was homogenized and the final volume was completed with Milli-Q water.

The next step involved the introduction of 55 g of $KMnO_4$ into a 100 mL beaker, the addition of ~ 800 mL of Millli-Q water and the mixture shaking during 2 hours. Then, a mixture of 2.75 g of ammonium oxalate and $KMnO_4$ was put into the shaker and kept in rest overnight (about 12 hours), covered with a polyethylene film and wrapped with an aluminum sheet. In the following day, the solution was kept in rest by 2 hours, its lower portion was disregarded and the remaining part was stored in a dark flask for avoiding the light exposition due to oxidation. Then, aqua regia ($HCl:HNO_3$ 3:1) was utilized for digesting the samples, which had been prepared during the chemical treatment of the sediments.

The calibration process consisted on the detection of the Hg content from solutions containing known amounts of this element (analytical standards), which varied from 2 to 40 μg/L and were prepared from a 1000 μg/L (1 mg/L = 1 ppm) Hg stock solution in an acid (5% HNO_3) and oxidizing (0.01% $K_2Cr_2O_7$) medium that was initially stored in a 250 mL volumetric flask and had been filled with 5% HNO_3. The lowest Hg concentration was determined by the equation:

$$2\ \mu g/L \times 250\ mL / 1000\ \mu g/L = 0.5\ mL$$

The others standards were prepared similarly on using crescent volumes from the stock solution (1000 μg/L). The construction of the calibration curve was automatically performed by the FIMS-400 atomic absorption spectrophotometer through the least squares statistical method. The dissolution process begins after the calibration curve construction, as schematically shown in figure 6.

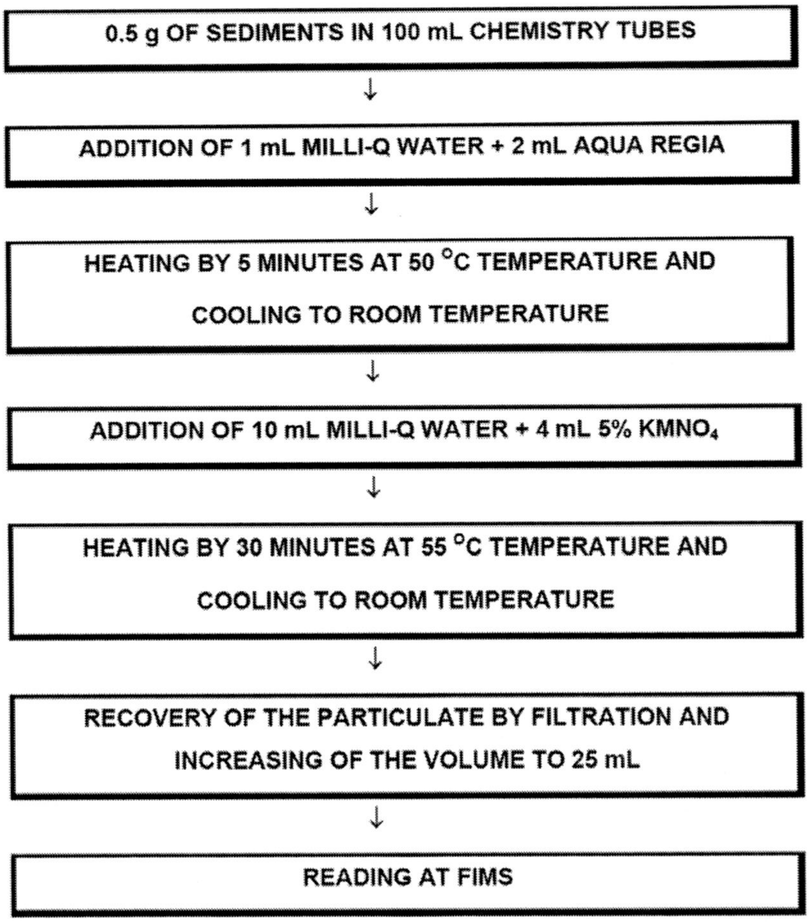

Figure 6. Schematic flow diagram illustrating the digestion process for sediments, soils and rocks aiming the determination of the total Hg content.

Table 4. Comparison of the Hg measurements performed at laboratories of the universities of Rio de Janeiro and Rondônia States in Brazil for several standards of sediments

STANDARD		UFRJ[1]			UNIR[2]			|AV1 - AV2|		t_{calc}[8]
No.	Code	AV1[3]	SD1[4]	FSD1[5] (%)	AV2[3]	SD2[4]	FSD2[5] (%)	VALUE[6]	(%)[7]	
01	IAEA356	6.470	0.493	8	6.010	0.180	3	0.460	7	2.48
02	OMSL167	0.210	0.037	18	0.230	0.010	4	0.020	10	1.48
03	OMSL169	1.270	0.020	2	1.580	0.050	3	0.310	24	16.28
04	PO2A	0.100	0.011	11	0.073	0.005	2	0.027	27	6.32
05	PO2B	0.076	0.006	8	0.076	0.006	8	0.000	0	0.00
06	PO2C	0.086	0.006	7	0.070	0.010	14	0.016	19	3.88
07	PO3A	0.086	0.006	7	0.086	0.006	7	0.000	0	0.00
08	PO3B	0.083	0.006	7	0.090	0.010	3	0.007	8	1.70
09	PO3C	0.560	0.440	79	0.070	0.011	16	0.490	87	3.15
10	PO4A	0.060	0.020	33	0.065	0.009	14	0.005	8	0.64
11	PO4B	0.050	-	-	0.067	0.003	4	0.017	34	-
12	PO4C	0.073	0.006	8	0.076	0.006	8	0.003	4	1.00
13	RASL1718	0.308	0.015	5	0.300	0.020	7	0.008	3	0.90
14	ROSD708	7.500	0.519	7	6.370	1.020	16	1.130	15	2.79
15	ROSL292	0.440	0.062	14	0.099	0.008	8	0.341	78	15.43
16	ROSL293	0.550	0.103	19	0.307	0.024	8	0.243	44	6.50
17	ROSL708	5.690	1.484	26	4.740	0.120	2	0.950	17	1.80
18	ROSL711	0.440	0.071	16	0.100	0.006	6	0.340	77	13.50
19	ROSL714	1.120	0.010	1	1.025	0.014	1	0.095	8	15.62
20	ROSL717	0.170	0.017	10	0.120	0.009	8	0.050	29	7.35
21	ROSL755	0.320	0.100	31	0.120	0.009	8	0.200	62	5.63
22	RRSL280	0.172	0.013	8	0.150	0.002	1	0.022	13	4.73
23	S1POUND	1.140	0.020	2	1.010	0.010	1	0.130	11	16.44
24	S2POUND	1.030	0.077	7	0.980	0.009	9	0.050	5	1.19

[1] UFRJ = Institute of Biophysics / Federal University of Rio de Janeiro;
[2] UNIR = Laboratory of Environmental Geochemistry / Federal University of Rondônia State;
[3] Average Hg content value (in $\mu g.g^{-1}$) for the triplicate measurements performed for each of the 3 aliquots of the standard, its
[4] standard deviation (SD), and the corresponding
[5] fractional standard deviation (FSD, in percentage);
[6] Module of the difference of the average Hg content values (AV1-AV2) and its
[7] corresponding percent value; [8] t_{calc} = t-calculated values (equation in the text).

All the analytical determinations at both laboratories (UFRJ and UNIR) were made in triplicate for the three sub-aliquots, yielding, therefore, nine measurements for each standard. Table 4 reports the values obtained at the laboratories, i.e. the average total Hg content, its standard deviation, and the fractional standard deviation. The values of the Hg content measured at UFRJ ranged from 0.05 up to 7.5 $\mu g.g^{-1}$ (average = 1.12 μg/g) whereas those obtained at UNIR varied between 0.06 and 6.4 $\mu g.g^{-1}$ (average = 0.99 μg/g).

If the errors obtained by both laboratories on the nine measurements performed for each standard are assumed to be normally distributed, then, it is possible to utilize the *t* test (statistical distribution of Student) for comparing the two average values. The reference value of the *t* test, t_{tab}, is 2.12 at 5% significance level and 16 degrees of freedom (Fisher and Yates, 1953), being represented in table 4 the t-values calculated according to following equation:

$$t_{calc} = \mid AV1 - AV2 \mid \times [(SD1^2 + SD2^2) / (N-1)]^{-1/2}$$

where N is the number of readings (N= 9) performed for each standard.

The null hypothesis that the average values obtained by both laboratories obey the same normal distribution is verified when $t_{calc} < t_{tab}$, situation observed in almost 50% of the cases reported in table 4. Such results suggest that discrepancies exist between the data supplied by both laboratories, which are related to random errors, not implying that the methods are incongruous.

The total Hg content for four standards of sediments evaluated by both laboratories had been previously determined by Malm et al. (1989), and also by the Laboratory of the GKSS Research Institute Geesthacht in Germany. Table 5 summarizes all the data obtained for the selected standards and figure 7 shows that excellent linear relationships (r^2 = 0.999) were found among the measured values by the different laboratories and investigators, demonstrating the effectiveness of the methodology settled at the Laboratory of Environmental Geochemistry of the Federal University of Rondônia State (UNIR).

Table 5. Comparison of the Hg measurements (in $\mu g.g^{-1}$) performed by different laboratories and investigators for some standards of sediments

STANDARD CODE	UFRJ1[1]	UFRJ2[2]	GKSS[3]	UNIR[4]	DIF(%)[5]
RRSL280	0.210	0.172	0.180	0.150	17
RASL1718	0.440	0.308	0.400	0.300	25
OMSL167	0.290	0.210	0.230	0.230	0
OMSL169	1.990	1.270	1.740	1.580	9
SPOUND[6]	-	1.085	-	0.995	-

[1] Data reported by Malm et al. (1989) and obtained at Institute of Biophysics, Federal University of Rio de Janeiro, Brazil;

[2] Unpublished data obtained by Bastos (this paper) at Institute of Biophysics, Federal University of Rio de Janeiro, Brazil;

[3] Data obtained at the Laboratory of the GKSS Research Institute Geesthacht, Germany;

[4] Data obtained at the Laboratory of Environmental Geochemistry, Federal University of Rondônia State, Brazil;

[5] Difference of the Hg content values (in percent) measured by the laboratories of GKSS and UNIR;

[6] Mean of the values reported in table 4 for standards Nos. 23 and 24.

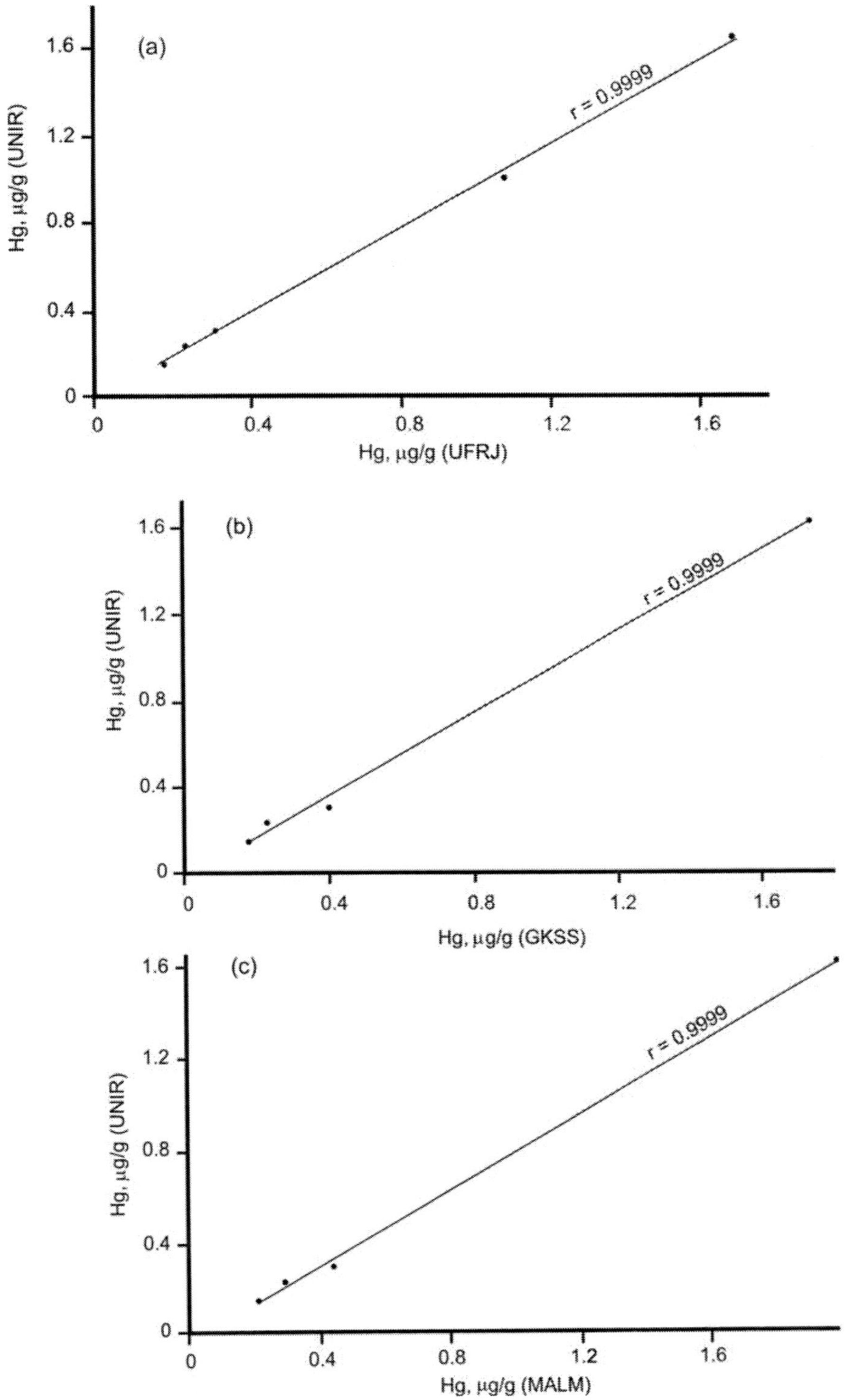

Figure 7. The relationships between the total Hg contents measured for some standards at a) Laboratory of Environmental Geochemistry-UNIR and Institute of Biophysics-UFRJ, b) Laboratory of Environmental Geochemistry-UNIR and Laboratory of the GKSS Research Institute, and c) Laboratory of Environmental Geochemistry-UNIR and Malm et al. (1989).

Chapter 5

MERCURY AND OTHERS PARAMETERS IN WATERS OF MADEIRA RIVER

The total drainage area of Madeira River corresponds to about 1.3×10^6 km^2. The surface area between Teotônio and Santo Antônio falls is around 260 km^2, including the urban area of Porto Velho municipality, the capital of Rondönia State. This chapter describes the results of chemical analyses performed for waters sampled along that sector of Madeira River.

SAMPLING

Water samples were collected in 18 points in the Madeira River between Teotônio and Santo Antônio waterfalls upstream from Porto Velho (area of about 260 km^2), along 25 km (figure 8); this sector of the river is not suitable for navigation. The fieldwork was performed from a small boat during the dry season, in a two-week period in August 1996. The water samples were manually collected from the channel or from the bank, depending on its access. They were stored in 15 liters polyethylene bottles that had been previously washed with Milli-Q water. The glassware and others flasks were also subjected to the same cleaning procedure that included washing with neutral detergent (Extran from Merck) into an acid solution, rinsing with Milli-Q water and drying in a stove. The flasks were rinsed in the field with the water being sampled, and, after collection, they were sealed with a PVC film and transported up to the laboratory.

ANALYTICAL PROTOCOL

Physical and physical-chemical parameters like temperature, pH and redox potential (Eh) were evaluated in the field. The pH measurement was performed by a digital portable Digimed digital meter coupled to a combination Analion glass electrode (model V-62-0-A); buffer solutions equilibrated with the sample temperature were utilized to calibrate the equipment before the analyses.

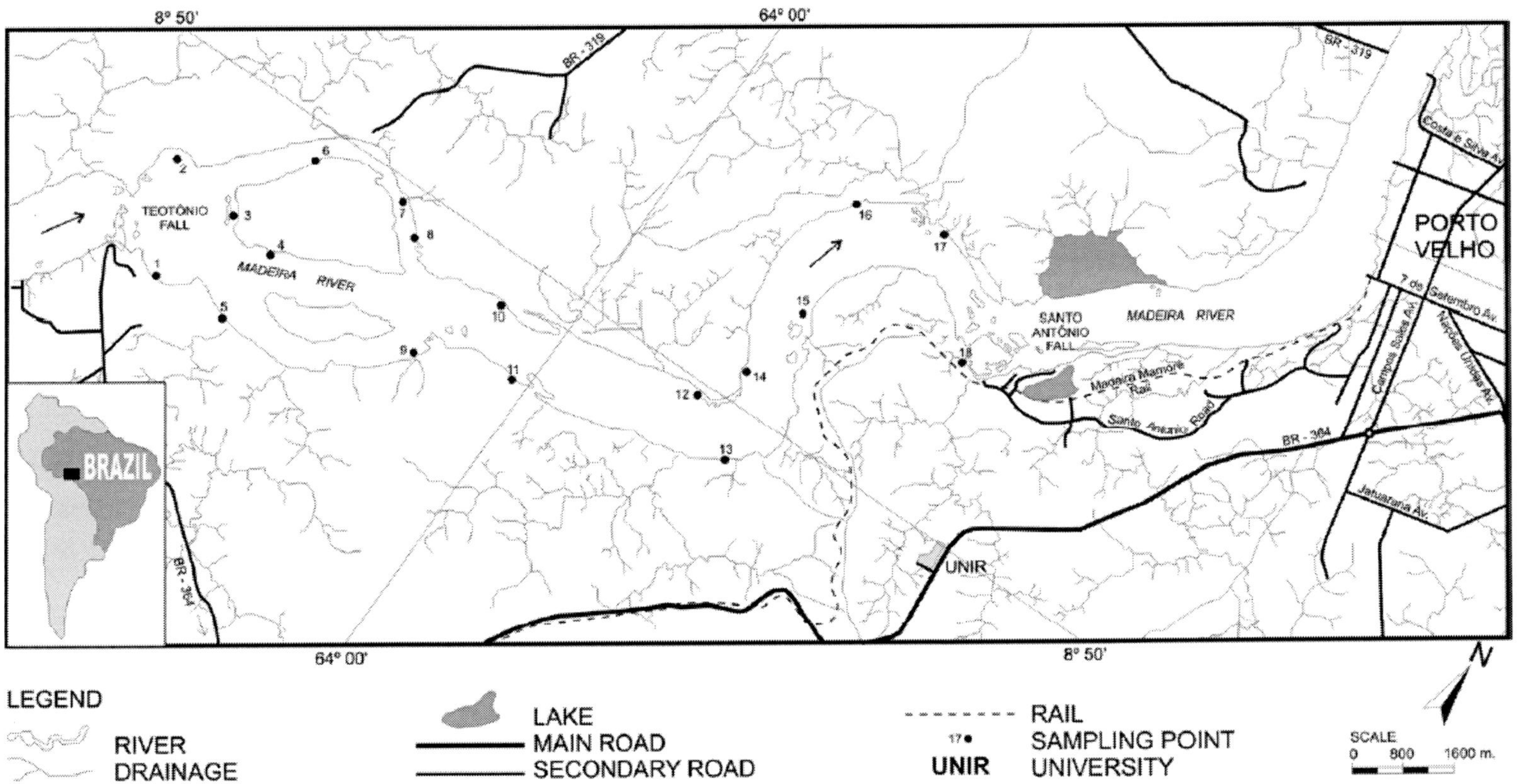

Figure 8. Location of Porto Velho city and sampling points for chemical analyses of waters, bottom sediments and suspended solids in Madeira River between Teotônio and Santo Antônio falls.

The Eh measurements in the water samples were realized with the same pH digital portable meter after disconnecting the pH electrode, selecting the "mV" position, and connecting a combination platinum electrode-Ag/AgCl reference element. Such electrode was previously calibrated on using a prepared Zobell I reference solution (Zobell, 1946) and adding the measured cell potential to the potential of the Ag/AgCl reference electrode at the temperature of measurement.

The difference between the measured Eh and the actual Eh of the reference solution was recorded and used to adjust the field readings, which were performed placing the electrode into an in-line flow cell. The sample readings were corrected by addition of the Ag/AgCl reference potential at the sample temperature and the calibration correction.

The sampling flasks were kept in iceboxes and transported up to the laboratory, where aliquots were divided for evaluating the suspended solids and characterizing again the pH by the same equipment that been utilized in the field. Different laboratories were engaged in the water analyses: Analytical Laboratory from CAERD (*Companhia de Águas e Esgoto* from Rondônia State, Porto Velho), Analytical Chemistry Laboratory from CENA (*Centro de Energia Nuclear na Agricultura* from USP-University of São Paulo, Piracicaba), and LABIDRO (*Laboratório de Isótopos e Hidroquímica* from UNESP-University of the State of São Paulo, Rio Claro).

The suspended solids were separated on filtering each sample through a 47-mm diameter Millipore membrane of 0.45-µm porosity. The total dissolved solids (TDS) content was evaluated on evaporating the filtrate to dryness in a weighed flask that was dried to constant weight at 180°C, with the increase in flask weight representing the TDS (APHA, 1989).

The true color of the water samples was determined by colorimetry (wavelength 455 nm) after filtering the suspended materials and using a program stored in the spectrophotometer that was calibrated in color units based on the APHA-recommended standard of 1 color unit being equal to 1 mgL^{-1} platinum as chloroplatinate ion (Hach, 1992).

The chemical oxygen demand (COD) in the water samples was measured by the dichromate reflux method (APHA, 1989), using aliquots acidified to pH=2 with concentrated H_2SO_4. Chloride was measured potentiometrically. Nitrate, nitrite and sulfate were determined by spectrophotometry (Hach, 1992).

The total alkalinity of the water samples was determined by titration using a titrator with sulfuric acid standard solution to an end point evidenced by the color change of a standard indicator solution (APHA, 1989). The obtained values corresponded to the bicarbonate concentrations since neither carbonate nor hydroxide were characterized. Diagrams and nomographs based on equations relating the ionization equilibria of the carbonates and water (APHA, 1989) enabled to evaluate the free carbon dioxide content of the samples from the following parameters measured in the laboratory: temperature, pH, dry residue (approximately total dissolved solids), and bicarbonate alkalinity.

The analyses of dissolved sodium and potassium were done using an atomic absorption spectrophotometer, whereas standard procedures were utilized for characterizing dissolved calcium, magnesium, silicon, iron, manganese, copper, zinc, lead, nickel, chromium, cadmium, barium, boron, and phosphorus in the liquid phase by inductively coupled plasma atomic emission spectrometry (ICP-AES).

Calcium hardness (as $CaCO_3$) and magnesium hardness (as $MgCO_3$) of the water samples were determined by the colorimetric method (wavelength 522 nm) after chelating calcium with EGTA and calcium and magnesium with EDTA (Hach, 1992), parameters that allowed to evaluate Ca and Mg contents lower than those obtained by ICP-AES, as well as the total hardness (TH), according to the equation (Todd, 1959): TH (mgL^{-1}) = 2.5 Ca (as $CaCO_3$) + 4.1 Mg (as $CaCO_3$).

RESULTS

Tables 6-8 report all results obtained in the liquid phase analysis, where the Ca and Mg contents are the values obtained by ICP-AES.

Charge imbalance ranged from 1.2% to 7.7%, being calculated as twice the charge imbalance normally computed, since the cation/anion imbalance was compared with the average of the cations and anions rather than with their sum (Ball and Nordstrom, 1992). Thus, the obtained values are not higher than 10%, generally indicative of errors in the analytical input values (Ball and Nordstrom, 1992).

The temperature of the waters is practically constant (30°C), their odor is absent, and the samples were slightly turbid, as indicated by their color, which was always higher than the value of 15 Pt-Co units (the maximum suggested by WHO (1984) for drinking-water quality). The amount of oxygen chemically consumed may be related to the presence of dissolved metals, perhaps iron. The field pH ranged from 6.1 to 6.8 (average = 6.5), indicating conditions tending to neutrality. The corrected Eh ranged from 141 to 238 mV (average = 156 mV), whereas the chemical oxygen demand (average = 4.6 mg/L) and free CO_2 (values generally lower than 1 mg/L) were practically constant.

The measured values for several dissolved elements in water samples were lower than the detection limit of the ICP-AES method, which was not accurate enough to provide the concentration data. The elements and respective detection limits are: Mn (0.02 mgL^{-1}), Cu (0.05 mgL^{-1}), Zn (0.05 mgL^{-1}), Pb (0.1 mgL^{-1}), Ni (0.05 mgL^{-1}), Cr (0.05 mgL^{-1}), Cd (0.05 mgL^{-1}), Ba (0.05 mgL^{-1}), B (0.05 mgL^{-1}), and P (0.1 mgL^{-1}).

The average values found for calcium (10 mg/L), magnesium (3.8 mg/L), silicon (4.8 mg/L), sodium = 2.8 mg/L) and potassium (1.6 mg/L) are not significantly different from those reported by Santos and Ribeiro (1988) for Madeira River, i.e.: calcium = 7.8 mg/L, magnesium = 1.5 mg/L, sodium = 6.8 mg/L, potassium = 3.6 mg/L and silicon = 5.6 mg/L. The major factors that have been considered responsible by the presence of magnesium, sodium and potassium in Amazonian waters are the weathering of feldspars, the pronounced lixiviation due to the high rainfall, the washing of leaves, stalks and trunks, and the organic decomposition (Schimidt, 1972; Tancredi et al., 1987).

Santos et al. (1981) considered the possibility of chloride inputs in hydrographic basins by runoff and re-mineralization of substances from the forests. Several of the elements analyzed exhibited concentration values lower than the detection limit, except iron and silicon (table 8). Their presence in waters from Madeira River is probably related to the dissolution secondary silicatic minerals and laterites that are abundant in the region, as a consequence of the weathering of hornblende, pyroxenes and biotite. The iron and Si^{+4} may also occur due to the presence of weathered feldspars and quartz in those rocks.

Mercury is present in the inorganic form in surface and ground waters at concentrations usually of less than 0.5 μgL^{-1} (WHO, 1993) despite the solubility evaluated in the past corresponding to 25 μgL^{-1} (Hem, 1970). Table 9 reports the dissolved Hg content in Madeira River water and its tributaries between the towns of Guajará Mirim near the Bolivian border and Porto Velho, which was determined by Nriagu et al. (1992) probably at the same hydrological period, due to the facilities for sampling.

Table 6. Results of the analyses of waters from Madeira River between Teotônio and Santo Antônio falls

Sample	Aspect	Color	Odor	Temp.	pH	pH	Eh (meas.)	Eh (correct.)	COD[1]	Free CO_2
No.		(PtCo)		(ºC)	(lab.)	(field)	(mV)	(mV)	(mg/L)	(mg/L)
1	slig. turbid	60	absent	31	7.85	6.53	226	153	4.2	< 1
2	slig. turbid	60	absent	30	8.02	6.28	227	155	4.8	< 1
3	slig. turbid	30	absent	31	7.78	6.51	214	141	4.4	< 1
4	slig. turbid	70	absent	30	7.90	6.13	222	150	4.2	< 1
5	slig. turbid	60	absent	30	7.82	6.67	227	155	5.0	< 1
6	slig. turbid	60	absent	30	7.68	6.60	222	150	4.4	1.2
7	slig. turbid	60	absent	30	7.68	6.56	222	150	4.6	< 1
8	slig. turbid	60	absent	30	7.81	6.60	222	150	4.8	< 1
9	slig. turbid	60	absent	30	7.79	6.40	229	157	4.4	< 1
10	slig. turbid	60	absent	30	7.74	6.46	217	145	4.8	1
11	slig. turbid	60	absent	30	7.81	6.75	220	148	5.2	< 1
12	slig. turbid	60	absent	30	7.79	6.61	228	156	5.0	< 1
13	slig. turbid	60	absent	30	7.86	6.60	225	153	4.4	< 1
14	slig. turbid	60	absent	30	7.91	6.65	222	150	3.8	< 1
15	slig. turbid	30	absent	30	7.73	6.57	310	238	4.4	< 1
16	slig. turbid	60	absent	30	7.72	6.31	237	165	5.2	< 1
17	slig. turbid	30	absent	30	7.73	6.79	219	147	4.4	< 1
18	slig. turbid	30	absent	32	7.77	6.80	216	142	4.2	< 1

[1] COD = Chemical Oxygen Demand. Analyses performed by Analytical Laboratory from CAERD.

Table 7. Results of the analyses (in mg/L) of waters from Madeira River between Teotônio and Santo Antônio falls

Sample No.	Cl^-	SO_4^{2-}	NO_2^-	NO_3^-	Alkalinity (HCO_3^-)	K	Na	Ca	Mg	Hardness (Ca)	Hardness (Mg)	Total Hardness	TDS[1]
1	9.9	13.7	0.004	0.1	30	1.9	3.1	10.0	3.9	7.6	0.7	22	111
2	9.9	11.9	0.013	0.2	24	1.4	2.9	10.1	3.8	7.2	0.9	20	123
3	14.9	5.3	0.001	0.7	28	1.4	2.9	9.9	3.7	7.2	1.4	25	82
4	9.9	14.6	0.008	0.1	22	1.8	2.7	10.4	4.0	7.2	0.9	22	124
5	9.9	11.9	0.004	0.1	26	1.5	2.7	10.1	3.8	7.6	0.7	22	132
6	9.9	11.9	0.004	0.1	30	1.6	2.9	10.8	4.2	7.2	1.2	23	112
7	9.9	13.7	0.004	0.1	28	1.6	2.9	10.2	3.9	7.2	0.9	22	85
8	9.9	11.9	0.004	0.1	24	1.6	2.9	10.1	3.9	7.2	0.9	22	114
9	9.9	13.7	0.004	0.1	20	1.2	2.7	10.0	3.8	7.6	1.2	24	142
10	9.9	14.6	0.004	0.4	28	1.5	2.7	10.2	3.9	7.2	0.9	22	121
11	9.9	12.8	0.004	0.5	24	1.5	2.7	10.1	3.9	7.6	0.7	22	147
12	9.9	12.8	0.004	0.1	22	1.1	2.9	10.1	3.9	7.2	0.9	22	124
13	9.9	12.8	0.004	0.1	24	1.4	2.9	9.9	3.8	7.2	1.2	23	109
14	9.9	12.8	0.017	0.5	28	2.0	2.9	10.0	3.8	7.2	0.9	22	115
15	14.9	2.2	0.001	0.9	26	1.3	2.9	9.5	3.7	7.2	1.4	24	102
16	9.9	11.9	0.004	0.4	26	3.8	2.9	10.0	3.8	7.2	0.9	22	106
17	14.9	4.5	0.001	0.9	26	1.3	2.9	9.6	3.7	8.0	1.2	25	93
18	14.9	3.0	0.001	0.6	24	1.3	2.7	9.7	3.7	8.0	1.2	25	75

[1] TDS = Total Dissolved Solids. Analyses performed by Analytical Laboratory from CAERD and Analytical Chemistry Laboratory from CENA.

Table 8. Results of the analyses (in mg/L) of waters from Madeira River between Teotônio and Santo Antônio falls

Sample No.	Fe	Cu	Zn	Ba	Cd	Cr	Pb	B	Ni	Mn	Si	P
1	0.05	< 0.05	< 0.05	< 0.05	< 0.05	< 0.05	< 0.10	< 0.05	< 0.05	< 0.02	4.7	< 0.1
2	0.2	< 0.05	< 0.05	< 0.05	< 0.05	< 0.05	< 0.10	< 0.05	< 0.05	< 0.02	4.9	< 0.1
3	< 0.05	< 0.05	< 0.05	< 0.05	< 0.05	< 0.05	< 0.10	< 0.05	< 0.05	< 0.02	4.7	< 0.1
4	< 0.05	< 0.05	< 0.05	< 0.05	< 0.05	< 0.05	< 0.10	< 0.05	< 0.05	< 0.02	4.8	< 0.1
5	0.09	< 0.05	< 0.05	< 0.05	< 0.05	< 0.05	< 0.10	< 0.05	< 0.05	< 0.02	4.8	0.1
6	0.1	< 0.05	< 0.05	< 0.05	< 0.05	< 0.05	< 0.10	< 0.05	< 0.05	< 0.02	5.0	< 0.1
7	0.2	< 0.05	< 0.05	< 0.05	< 0.05	< 0.05	< 0.10	< 0.05	< 0.05	< 0.02	5.0	< 0.1
8	0.2	< 0.05	< 0.05	< 0.05	< 0.05	< 0.05	< 0.10	< 0.05	< 0.05	< 0.02	4.9	< 0.1
9	0.2	< 0.05	< 0.05	< 0.05	< 0.05	< 0.05	< 0.10	< 0.05	< 0.05	< 0.02	4.9	0.1
10	0.1	< 0.05	< 0.05	< 0.05	< 0.05	< 0.05	< 0.10	< 0.05	< 0.05	< 0.02	4.9	0.1
11	0.2	< 0.05	< 0.05	< 0.05	< 0.05	< 0.05	< 0.10	< 0.05	< 0.05	< 0.02	4.9	< 0.1
12	0.1	< 0.05	< 0.05	< 0.05	< 0.05	< 0.05	< 0.10	< 0.05	< 0.05	< 0.02	4.8	0.1
13	0.09	< 0.05	< 0.05	< 0.05	< 0.05	< 0.05	< 0.10	< 0.05	< 0.05	< 0.02	4.7	0.1
14	0.1	< 0.05	< 0.05	< 0.05	< 0.05	< 0.05	< 0.10	< 0.05	< 0.05	< 0.02	4.8	0.1
15	0.1	< 0.05	< 0.05	< 0.05	< 0.05	< 0.05	< 0.10	< 0.05	< 0.05	< 0.02	4.8	< 0.1
16	0.08	< 0.05	< 0.05	< 0.05	< 0.05	< 0.05	< 0.10	< 0.05	< 0.05	< 0.02	4.8	< 0.1
17	< 0.05	< 0.05	< 0.05	< 0.05	< 0.05	< 0.05	< 0.10	< 0.05	< 0.05	< 0.02	4.7	< 0.1
18	0.2	< 0.05	< 0.05	< 0.05	< 0.05	< 0.05	< 0.10	< 0.05	< 0.05	< 0.02	4.8	< 0.1

Analyses performed by Analytical Chemistry Laboratory from CENA.

The sensitivity of the equipment for Hg measurements settled at the Environmental Geochemistry Laboratory from UNIR, Porto Velho, did not allow obtain data for dissolved and particulate phases. The results reported by Nriagu et al. (1992) in table 9 are useful for geochemical calculations, since they were generated through cold vapor atomic absorption spectrometry in an ultra-clean laboratory equipped with gold-coated air filters, after using scrupulous methods in the field to exclude sample contamination. The average value of 13.8 ngL^{-1} was about 17-fold higher than the mean dissolved Hg concentration in the rivers of the world (Nriagu et al., 1992), but 36-fold lower than the upper concentration of 0.5 μgL^{-1} established by WHO (1993).

Table 9. Mercury concentrations (in ngL^{-1}) in waters from Madeira River and tributaries

Sample No.	Location	Dissolved	Particulate	Total
1	Mutum Paraná (junction with Madeira River)	11.3	14.5	25.8
2	Mutum Paraná (30 m from highway)	9.8	17.5	27.3
3	Prainha	15.0	18.2	33.2
4	Humaitá, mid-river	15.8	6.4	22.2
5	Humaitá, nearshore	12.5	8.5	21.0
6	Porto Velho	13.7	10.4	24.1
7	Teotônio Falls	13.3	6.6	19.8
8	2 km upstream Teotônio Falls	10.7	9.1	19.9
9	4 km upstream Teotônio Falls	16.3	6.2	22.5
10	8 km upstream Teotônio Falls	17.2	8.5	25.6
11	10 km upstream Teotônio Falls	12.7	9.1	21.9
12	12 km upstream Teotônio Falls	17.1	15.3	32.4
Average of Samples No. 6-12		14.4	9.3	23.7
Total Average		13.8	10.8	24.6

The data are from Nriagu et al. (1992)

Chapter 6

HYDROCHEMICAL RELATIONSHIPS IN MADEIRA RIVER

The Madeira River is moderately turbid, mainly when occurs an increase of the particulate matter during the rainy season. This chapter focuses the major chemical features of the waters close to Porto Velho, which were identified from the results reported in the previous chapter.

THE USE OF THE PIPER DIAGRAM FOR CLASSIFYING THE WATERS

The Piper (1944) diagram may be properly utilized to identify the chemical facies of the waters from Madeira River between Teotônio and Santo Antônio falls. The cations (in mEq/L – percentage values) are represented by a point inside a triangle, whereas the anions (in mEq/L – percentage values) are represented by a point in another triangle. The two points are projected in a central diamond, resulting in an unique point that describes the total ionic distribution, i.e. the dominant chemical character of the water and the similarities and differences among several water samples (Bonotto and Mancini, 1992). The results reported in the last chapter have been plotted in figure 9. It is possible verify that the analyzed waters are Ca^{2+} dominated/mixed in terms of dissolved cations, whereas they are mixed in terms of dissolved anions.

THE USE OF THE GIBBS DIAGRAM FOR CLASSIFYING THE WATERS

Gibbs (1970) suggested a chemical classification of surface waters based on values of total dissolved solids and $Na^+/Na^+ + Ca^{2+}$ ratios. These parameters were utilized for constructing a boomerang-type diagram from data available for the major rivers worldwide (figure 10). The rivers strongly dominated by influence of meteoric waters (rainfall) are inserted in the lower portion of the diagram, whereas those mainly affected by evaporation and precipitation processes are inserted in the upper part of the diagram; the more central part of the diagram is occupied by rivers strongly dominated by weathering processes affecting the rocks. The $Na^+/ Na^+ + Ca^{2+}$ ratios in waters inserted in the borders of the boomerang diagram tend to reach the unit value. The Na^+ concentration in rainwater is higher than that of Ca^{2+} in the lower portion of the diagram, especially in coastal areas, and, consequently, the $Na^+/ Na^+ + Ca^{2+}$ ratio approaches unit. The $Na^+/ Na^+ + Ca^{2+}$ ratio also tends to reach unity at the upper portion of the diagram, as a consequence of calcite precipitation caused by evaporation processes, where occurs calcium depletion and sodium enrichment in the liquid phase. Thus, rivers from arid regions, saline lakes and oceans exhibit high $Na^+/ Na^+ + Ca^{2+}$ ratios whereas Amazonas River and tributaries are good examples of waters possessing an accentuated contribution of rainfall, as shown in figure 10.

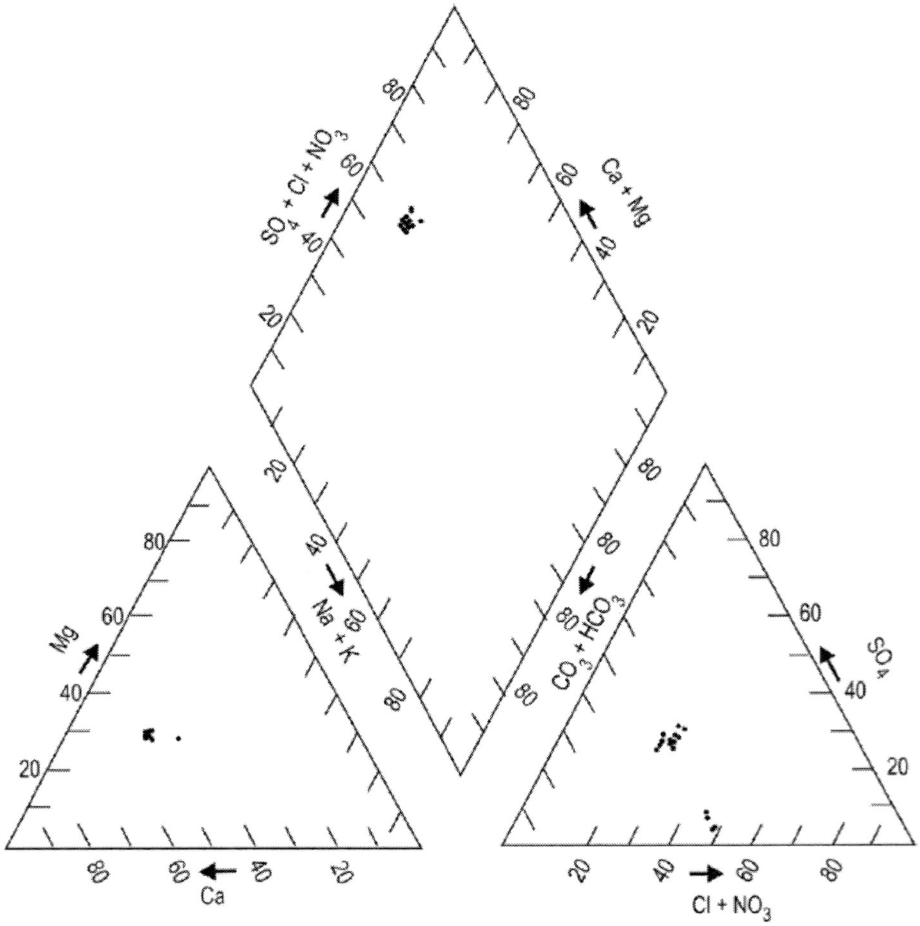

Figure 9. Data for surface waters from the studied area plotted on a typical Piper (1944) diagram.

The mean values of TDS (Total Dissolved Solids), sodium and calcium contents reported in the last chapter have been plotted in figure 10. It is possible verify that the $Na^+/ Na^+ + Ca^{2+}$ ratio is similar to that given by Gibbs (1970), however, the TDS value is higher, causing the insertion of the "average sample" in the central portion of the diagram rather than in its lower one. According to this scenario, the water chemistry would be characterized by less of an influence of dilution processes by rain than due to the enhanced effect of rock-soil weathering processes. The typical conditions of high rainfall and temperature in the area facilitate the occurrence of weathering processes, where the dip and smooth relief forming a vast drainage network are favorable to cause the dissolution of cations and anions. Such situation was also evidenced by others chemical analysis of the waters, because when the data are used to study stability relations of mineral phases in the systems K_2O-Al_2O_3-SiO_2-H_2O, CaO- Al_2O_3-SiO_2-H_2O and Na_2O-Al_2O_3-SiO_2-H_2O, it is found that all samples fall into the kaolinite field of stability (figure 11).

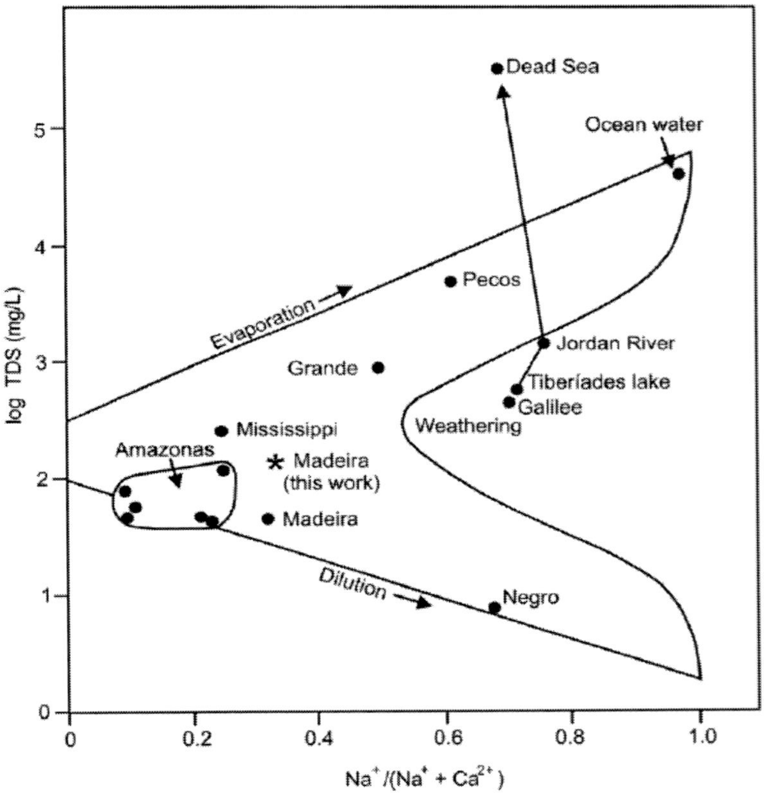

Figure 10. Average data for surface waters from the studied area plotted on a Gibbs (1970) diagram.

pH and Redox Potential Variations

The pH values measured in the field are lower than those evaluated in the lab (figure 12). The possible reasons may be the different sample agitation during pH readings and CO_2-loss when the samples had been transported up to the lab (Back and Hanshaw, 1965). The higher pH values measured in the laboratory and the low (up to 30 mgL^{-1}) abundance of bicarbonate in the waters implied on the low CO_2 values reported in table 6.

The field pH values are between 6.1 and 6.8, whereas the corrected Eh redox potential ranges from 142 to 238 mV. If the pH and Eh data obtained for the waters are inserted on a traditional Eh-pH diagram (Krauskopf, 1979; Krauskopf and Bird, 1995), then, it is possible identify a transitional field, tending to be reducing in character (figure 13). The low presence of dissolved oxygen (aeration) may be related to the sampling period, typically a dry one (August/1996).

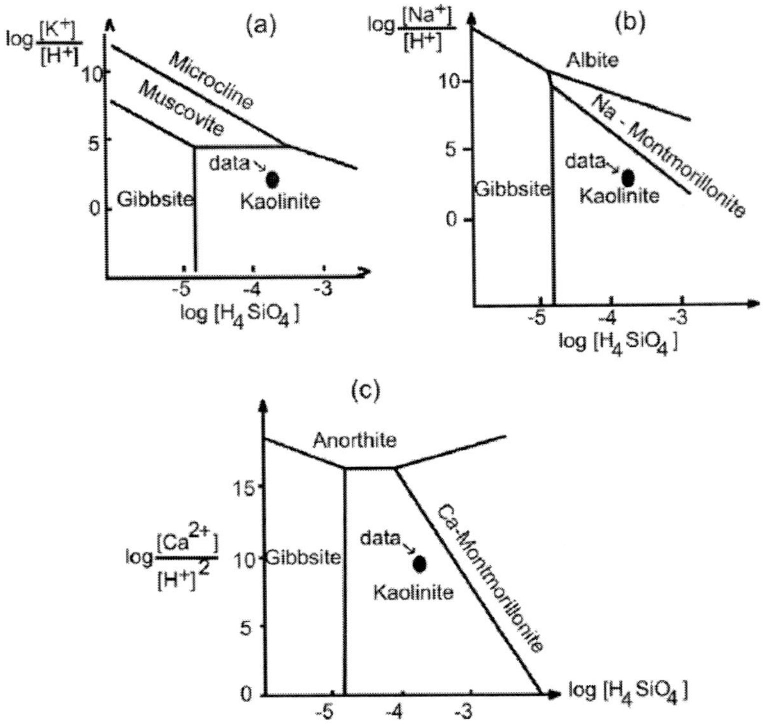

Figure 11. Chemical data for the studied waters plotted on diagrams representing the stability relations of some mineral phases in the systems a) K_2O-Al_2O_3-SiO_2-H_2O, b) Na_2O-Al_2O_3-SiO_2-H_2O, and c) CaO-Al_2O_3-SiO_2-H_2O at 25°C and 1 atmosphere (Garrels and Christ, 1965).

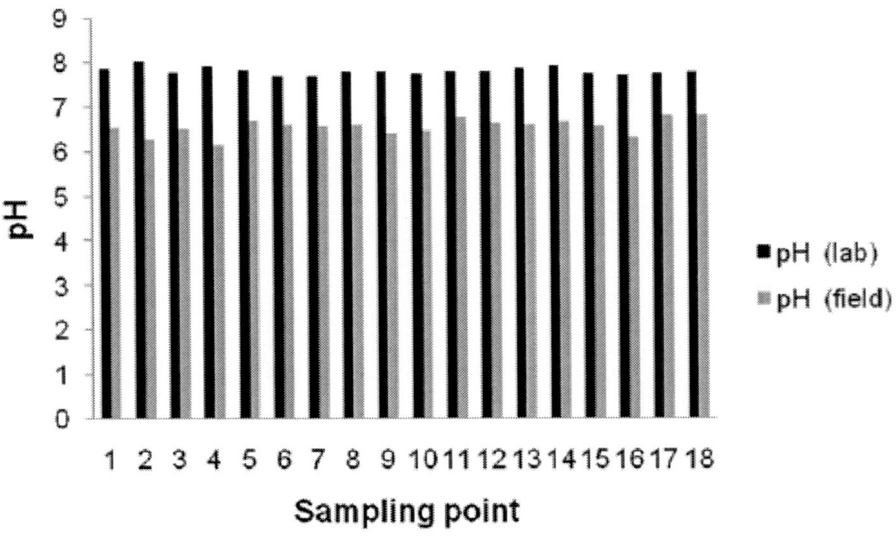

Figure 12. The pH variation for surface waters from the studied area.

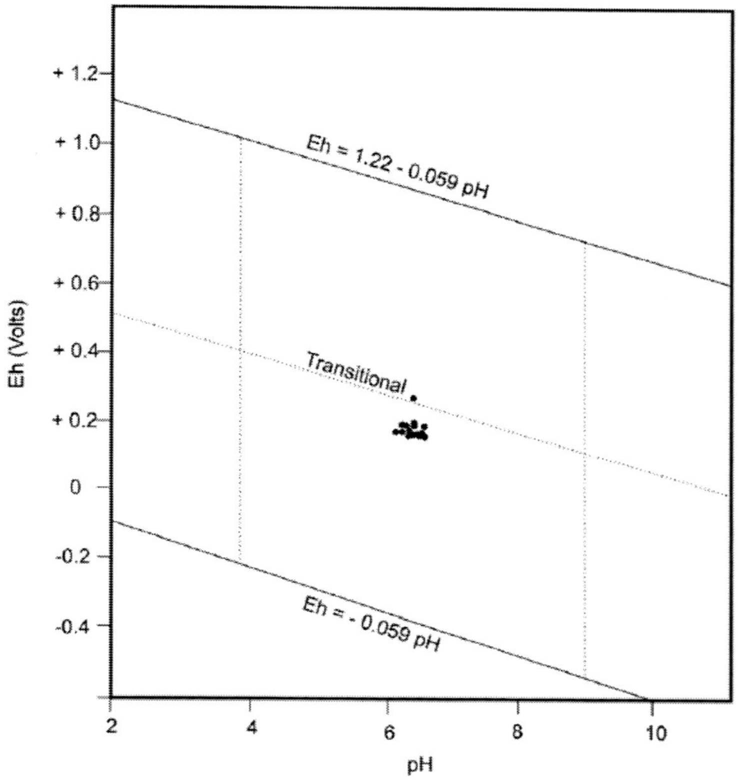

Figure 13. Chemical data for the studied waters plotted on an Eh-pH diagram at 25°C temperature and 1 atm partial pressure (Krauskopf, 1979).

Eh - pH Diagram for Mercury Compounds

Under acid conditions, mercury precipitates associated to humic and fulvic acids, exhibiting strong affinity by hydrated minerals, mainly clay minerals enriched in iron oxides and organic matter (Siegel and Siegel, 1975). Harriss (1978) demonstrated that some adsorption processes (specific or electrostatic) of mercury with clay minerals iron-enriched are extremely strong and fast, whereas desorption is very slow. This probably explains their long retention periods in aquatic environments and difficult remobilization by natural processes. However, such situation is not expected at the area studied as a consequence of the pH values tending to neutrality.

In aquatic environments containing aerated waters and Eh > 0.5 V, there is predominance of bivalent mercury species (Hg^{2+}) with strong covalent binding formed with chloride, under modeling conditions corresponding to a solution containing 36 mg/L of Cl^- and 96 mg/L of sulfate (Hem, 1970). However, in moderately reducing environments, mercury is more stable in the metallic (Hg^0) form, that is a condition favorable for its precipitation as sulfides (HgS) exhibiting low solubility. Björnberg et al. (1988) pointed out that the bioavailable mercury is directly related to the sulfides activity that depends on the pH conditions; in neutral and alkaline waters, it tends to be fixed by the formation of HgS.

The mercury in moderately reducing environment is influenced by the stability of the complex ions $HgCl_4^{2-}$ and HgS_2^{2-}, which, in solution, form cinnabar deposits from aqueous media at low temperature (epithermal deposits). Moore and Ramamoorthy (1984) admitted that the mercury sulfide (HgS) is extremely important on the Hg mobilization in less oxidizing and reducing environments as a consequence of the low solubility product (10^{-53}).

The results obtained for pH and redox potential in the waters from Madeira River between Teotônio and Santo Antônio falls were plotted in the Eh-pH diagram for metallic Hg, the sulfide cinnabar and the oxide montroydite in a system containing $\Sigma S = 10^{-3}$ mol.L^{-1} at 25°C and 1 atm (Faure, 1991) (figure 14). The data suggest that mercury may be present as the metallic Hg° inert form in the area, confirming the results of other investigators (Pfeiffer and Lacerda, 1988; Lacerda and Salomons, 1992). The eventual presence of mercury in the ionized state would imply on favorable conditions to form adsorption reactions with clay minerals, iron/manganese hydrated oxides or complex ions associated to chloride and/or methylmercury hydroxides. However, Frenet (1979) also considers possible the occurrence of metallic mercury adsorption to clay minerals as the ions $HgCl_4^-$, $HgCl_3^-$ or $HgCl^+$ based on the results of experiments performed with illite, kaolinite and mercurous chloride. Such proposal is reinforced by the identification of illite in bottom sediments occurring in the area studied.

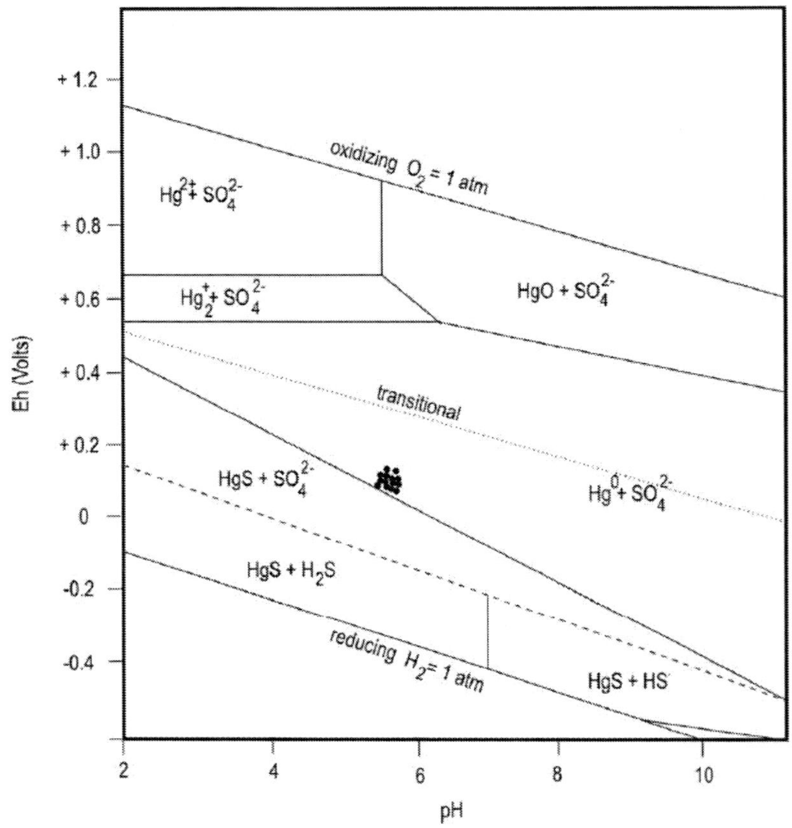

Figure 14. Data for surface waters from the studied area plotted on an Eh-pH diagram for metallic Hg, the sulfide cinnabar and the oxide montroydite in a system containing $\Sigma S = 10^{-3}$ mol.L^{-1} at 25°C and 1 atmosphere (Faure, 1991).

Chapter 7

MERCURY AND OTHERS PARAMETERS IN ROCKS, SUSPENDED SOLIDS AND SEDIMENTS FROM MADEIRA RIVER

The sediments are important sinks for accumulation of heavy metals, inclusive mercury. This chapter describes the results obtained in the chemical analyses of the samples of rocks, suspended solids and sediments collected between Teotônio and Santo Antônio falls in Madeira River.

SAMPLING

The bottom sediments were collected in the same 18 points in Madeira River utilized for water sampling, i.e. between Teotônio and Santo Antônio waterfalls, upstream from Porto Velho city (figure 8). About 500 g of sediments were collected from each sampling point with a bottom dredge (Petersen-type) slowly trawled upstream and transferred to a half-filled thick polyethylene bag, which was closed by adhesive tape after removing all air above the sample by tightly twisting their opening and folding the bag into a U shape.

All bags were stored in iceboxes and transported up to the laboratory, where the material was dried at 60°C, crushed, quartered, sieved for separation of the <200 mesh (<74 μm) size fraction, and divided into aliquots for performing X-ray diffraction analyses and measuring the total Hg, major cations, metals, boron, phosphorus, and organic matter.

The suspended solids in water were separated from the liquid phase through 0.45 μm Millipore membrane; the experimental steps were held at the Environmental Geochemistry Laboratory belonging to UNIR (*Fundação Universidade Federal de Rondônia*), Porto Velho city, Rondônia State. Digestion and extraction procedures described by Malm et al. (1989) were used for Hg analysis in the sediments. They were performed on a Perkin Elmer model FIMS-400 atomic absorption spectrophotometer. Organic matter content in the sediments has been estimated by colorimetry (Hach, 1992).

Sediments (1 g of the <74 μm size fraction) and the filters containing the particles were solubilized through acid digestion. Sodium and potassium were analyzed by atomic absorption technique, whereas calcium, magnesium, iron, manganese, copper, zinc, lead, nickel, chromium, cadmium, barium, boron, and phosphorus concentrations were obtained by ICP-AES method.

Six rock samples were collected at Teotônio and Santo Antônio waterfalls, where coarse-grained igneous rocks corresponding to granites, syenites, and monzonites have been characterized in terms of petrographic, geochemical, and geochronological aspects (Payolla, 1994; Verschure and Bon, 1972; Bettencourt et al., 1995). The Rb-Sr and U-Pb radiometric methods yielded ages for these rocks ranging from 1201 up to 1406 Ma (Verschure and Bon, 1972; Bettencourt et al., 1995). The following samples were collected at Teotônio fall: R1 (fayalite-chlinopyroxene-amphibole alkali-feldspar syenite), R2 (chlinopyroxene-amphibole alkali-feldspar granite) and R3 (fayalite-chlinopyroxene-amphibole quartz-syenite). The following samples were collected at Santo Antônio fall: R4 (biotite monzogranite), R5 (biotite monzogranite) and R6 (biotite monzogranite).

The rock fragments were size-reduced in two jaw crushers: a) Furlan BN-20 (openings-input: 5 cm and output: 1.5 cm); b) Furlan BN-20 (openings- input: 1.5 cm and output: 3-5 mm). Then, an additional size-reduction (< 400 mesh) was performed in an *oscillating mill* KHD - Industricanlagem AG Humboldt Wedag.

Analytical Protocol

The suspended solids separated on filtering each sample through a 47-mm diameter Millipore membrane of 0.45-μm porosity were weighted and dissolved according to the technique described by Caprioglio and Bonotto (1997). An amount of sediments corresponding to 1 g was inserted into a Parr-type digestion system. Then, a mixture of different acids was added, heated and evaporated to dryness. The process was repeated some times until complete dissolution of the material.

The Analytical Chemistry Laboratory from CENA (*Centro de Energia Nuclear na Agricultura* from USP-University of São Paulo, Piracicaba) performed the following chemical analyzes for bottom sediments and suspended solids: potassium, sodium, calcium, lead, iron, copper, zinc, barium, cadmium, chromium, magnesium, boron, nickel, silicon, manganese and phosphorus. The results for silicon correspond to minimum values due to the analytical procedure utilized for the sample dissolution. The organic matter analysis for bottom sediments was conducted at LABIDRO (*Laboratório de Isótopos e Hidroquímica* from UNESP-University of the State of São Paulo, Rio Claro) by spectrophotometry, according to the procedure described by Walkley (1947). The mineralogical characterization of the sediments was realized by X-ray diffraction at the Geochemistry Laboratory from UNESP-University of the State of São Paulo, Rio Claro.

Mercury in Rocks And Sediments

The results of the measurements of the total Hg content in bottom sediments and rocks collected between Teotônio and Santo Antônio falls are shown in table 10 and figure 15.

Table 10. Results of the measurements of the total Hg content in bottom sediments and rocks collected on the Madeira River, near Porto Velho city, Rondônia State

SAMPLE No.	MATERIAL	SITE OF SAMPLING[1]	Hg ($\mu g.g^{-1}$)	REMARKS
1	Sediments	Madeira River	0.0543	-
2	Sediments	Madeira River	0.0518	-
3	Sediments	Madeira River	0.1357	-
4	Sediments	Madeira River	0.0454	-
5	Sediments	Madeira River	0.0402	-
6	Sediments	Madeira River	NM[2]	-
7	Sediments	Madeira River	0.0444	-
8	Sediments	Madeira River	0.0209	-
9	Sediments	Madeira River	0.0456	-
10	Sediments	Madeira River	0.0583	-
11	Sediments	Madeira River	0.0391	-
12	Sediments	Madeira River	0.0481	-
13	Sediments	Madeira River	0.0480	-
14	Sediments	Madeira River	0.0576	-
15	Sediments	Madeira River	0.0408	-
16	Sediments	Madeira River	0.0604	-
17	Sediments	Madeira River	0.0365	-
18	Sediments	Madeira River	0.0520	-
R1	Rock	Teotônio waterfall	0.0072	fayalite-chlinopyroxene-amphibole alkali-feldspar syenite
R2	Rock	Teotônio waterfall	0.0037	chlinopyroxene-amphibole alkali-feldspar granite
R3	Rock	Teotônio waterfall	0.0040	fayalite-chlinopyroxene-amphibole quartz-syenite
R4	Rock	Santo Antônio waterfall	0.0036	biotite monzogranite
R5	Rock	Santo Antônio waterfall	0.0045	biotite monzogranite
R6	Rock	Santo Antônio waterfall	0.0033	biotite monzogranite

[1] All locations are shown in figure 8; [2]NM = not measured.

Fine sediments play a significant part in the transport and fate of mercury in tropical aquatic systems in the Amazon region, and table 10 shows that the Hg concentration in the bottom sediments ranged from 0.0209 up to 0.1357 $\mu g.g^{-1}$, average of 0.0517 $\mu g.g^{-1}$. Such values are extremely lower than those varying between 65 and 1675 $\mu g.g^{-1}$ for bottom sediments from Teotônio waterfalls, which were evaluated by Pfeiffer et al. (1989) during the peak gold mining activities that involved the use of a large number (7,000 – 8,000) of mechanical dredges operating simultaneously. Values as well reasonably high, ranging from 0.21 up to 19 $\mu g.g^{-1}$, were also reported by Pfeiffer et al. (1989) for the Hg content in bottom sediments from Mutum Paraná River, a tributary of the Madeira River. However, the values obtained in this investigation don't differ very significantly from those reported by Lacerda et al. (1987) for bottom sediments of the Madeira River (ranging from 0.05 up to 0.28 $\mu g.g^{-1}$)

and by Martinelli et al. (1988) for bottom sediments of the Madeira River between the towns of Porto Velho and Uricurituba (Amazonas State) (ranging from 0.03 up to 0.35 μg.g^{-1}).

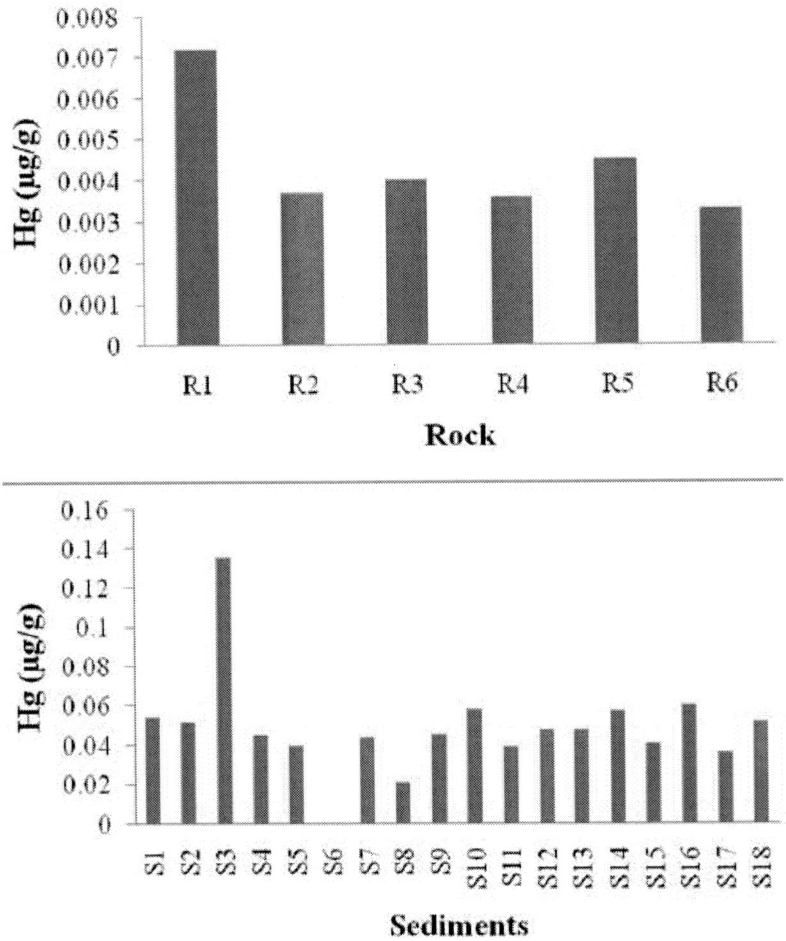

Figure 15. The total Hg content values measured for samples of rocks and bottom sediments collected near Porto Velho city, Rondônia State.

Table 10 shows that the Hg concentration in the rocks varied between 0.0033 and 0.0072 μg.g^{-1}, average of 0.0044 μg.g^{-1}. The values ranged from 0.0037 to 0.0072 μg/g (mean = 0.005 μg/g) for rocks from Teotônio fall, whereas from 0.0033 to 0.0045 μg/g (mean = 0.004 μg/g) for biotite monzogranites from Santo Antônio fall. These results provide guideline background values for mercury concentration in the area comprised between Teotônio and Santo Antônio falls, which can be compared with data from others regions. For instance, Jovanovic and Reed (1968) and Jonasson and Boyle (1979) reported total Hg contents for several rock types, for instance, between 0.007 and 0.25 μg.g^{-1} for ultrabasics, between 0.005 and 0.084 μg.g^{-1} for basics, and between 0.03 and 0.09 μg.g^{-1} for quartzites and amphibolites. The mean total Hg concentration in the rocks from Teotônio and Santo Antônio waterfalls is lower than these referred values for metamorphic rocks, being also lower than the values inserted on the range of 0.007 up to 0.2 μg.g^{-1} reported by Lindqvist et al. (1984) for acid rocks. Therefore, the obtained values for the Hg content in the rock samples from Teotônio and Santo Antônio waterfalls may be considered as appropriate for defining the regional background, since they represent analyses of samples free from anthropogenic inputs of Hg due to the processing of alluvial gold ores.

Therefore, the total Hg content in bottom sediments collected near the town of Porto Velho is close to other reported values in the area, but, in spite of this, the mean value corresponding to 0.0517 $\mu g.g^{-1}$ is about 10 times higher than the average of 0.0044 $\mu g.g^{-1}$ found for the analyzed rocks (regional background). Intense weathering processes affecting the crystalline rocks outcropping at the studied area have been partially responsible for the generation of fine sediments transported by water and deposited at Madeira River, and, due to this reason, it would be expected that similar Hg concentrations would be obtained in the analyzed rocks and bottom sediments. Because this was not the case, it is possible to suggest that the influence of the Hg anthropogenic emissions that occurred in the past during the accentuated gold-mining activities on the Madeira River still remains in the present days, when such activities were reduced substantially, being practically absent.

The sensitivity of the equipment for Hg measurements settled at the Environmental Geochemistry Laboratory from UNIR, Porto Velho, did not allow obtain Hg data for the suspended solids (particulate phase). The results reported by Nriagu et al. (1992) in table 9 are useful for geochemical calculations. The values ranged from 6.2 to 18.2 ng/L, and the average value (10.8 ng/L) is slightly lower than the mean value of 13.8 ng/L reported in the liquid phase (table 9).

Chemical Analysis in Sediments

Table 11 reports the results of the chemical analyses of bottom sediments from Madeira River between Teotônio and Santo Antônio falls. The average values are: potassium = 14,065 μg/g, sodium = 4,558 μg/g, iron = 38,101 μg/g, copper = 636 μg/g, zinc = 421 μg/g, barium = 103 μg/g, cadmium = 5 μg/g, chromium = 5 μg/g, lead = 169 μg/g, calcium = 3,025 μg/g, magnesium = 6,205 μg/g, boron = 13 μg/g, nickel = 54 μg/g, manganese = 582 μg/g and phosphorus = 951 μg/g. In the case of copper and zinc, the average was calculated disregarding samples 9 and 13 that exhibit concentration values much higher than the others, probably due to contamination (in the field or laboratory).

The data in table 11 indicate that the most abundant elements/compounds in the bottom sediments are: organic matter, iron, potassium, magnesium, sodium and calcium. The values obtained for silicon ranged from 20 to 60 μg/g and they are minima as the samples were digested with HF; consequently, no Si data are reported in table 11.

Chemical Analysis in Suspended Solids

Table 12 reports the results of the amount of suspended solids from Madeira River between Teotônio and Santo Antônio falls. The particulate matter is relatively high, ranging from 11.8 to 39.03 mg/L and agreeing with the turbid character of those waters. Table 13

Table 11. Results of the analyses (in μg/g) of bottom sediments from Madeira River between Teotônio and Santo Antônio falls

Sample No.	K	Na	Fe	Cu	Zn	Ba	Cd	Cr	Pb	Ca	Mg	B	Ni	Mn	P	Organic Matter (%)
1	13,950	4,690	39,140	90	190	350	<5	5	150	3,090	6,190	10	50	610	960	1.15%
2	12,190	4,630	41,870	520	530	100	<5	<5	130	3,310	5,250	10	50	730	940	0.80%
3	14,670	5,000	38,990	80	130	80	<5	5	150	3,140	6,130	10	50	680	940	1.03%
4	13,590	4,480	35,210	860	420	100	<5	5	120	2,680	6,040	8	40	520	85	1.29%
5	15,040	4,790	38,790	240	250	100	<5	5	150	3,280	6,670	60	50	610	1,000	1.13%
6[1]	-	-	-	-	-	-	-	-	-	-	-	-	-	-	-	-
7	13,060	4,690	35,480	310	260	60	<5	<5	130	2,950	5,770	9	50	550	900	0.98%
8[1]	-	-	-	-	-	-	-	-	-	-	-	-	-	-	-	-
9	12,110	4,380	31,460	19,750	11,000	100	5	<5	360	2,780	5,570	8	70	430	1,430	1.08%
10	14,400	4,580	40,440	220	220	60	<5	5	160	3,360	6,670	10	50	600	1,010	1.28%
11	11,850	4,900	28,800	1,220	830	80	<5	<5	110	2,670	4,870	8	40	390	760	0.61%
12[1]	-	-	-	-	-	-	-	-	-	-	-	-	-	-	-	-
13	14,310	4,480	39,670	9,070	13,850	100	5	5	230	3,090	6,340	10	80	570	1,270	1.40%
14[1]	-	-	-	-	-	-	-	-	-	-	-	-	-	-	-	-
15	17,700	4,580	48,670	90	150	120	<5	6	180	3,200	7,520	10	60	680	1,150	1.75%
16	13,590	3,970	34,720	2,520	620	60	<5	5	160	3,030	6,690	8	60	550	970	1.51%
17	14,770	4,580	37,680	110	670	60	<5	5	140	2,640	6,130	10	50	560	890	1.13%
18	15,690	4,070	42,500	1,370	780	80	<5	5	200	3,140	7,030	10	50	670	1,020	1.61%

[1] Not Measured. Analyses performed by Analytical Chemistry Laboratory from CENA and LABIDRO from IGCE-UNESP.

Table 12. Results of the evaluation of the suspended solids amount in Madeira River between Teotônio and Santo Antônio falls

Sample No.	FILTER WEIGHT[1] (g)	FILTER WEIGHT[2] (g)	AMOUNT RETAINED (g)	FILTERED VOLUME (L)	FINAL VALUE (mg/L)
1	0.0492	0.0577	0.0085	0.720	11.80
2	0.0657	0.0867	0.0210	0.920	22.83
3	0.0683	0.0848	0.0165	1.180	13.98
4	0.1028	0.1233	0.0205	0.900	22.78
5	0.0645	0.0794	0.0149	1.000	14.90
6	0.0637	0.0920	0.0283	0.725	39.03
7	0.0669	0.0894	0.0225	0.780	28.85
8	0.0623	0.0762	0.0139	0.650	21.38
9	0.0635	0.0756	0.0121	0.830	14.58
10	0.0655	0.0824	0.0169	0.920	18.37
11	0.0663	0.0769	0.0106	0.850	12.47
12	0.0670	0.0842	0.0172	0.900	19.11
13	0.0668	0.0823	0.0155	0.835	18.56
14	0.0635	0.0764	0.0129	0.850	15.17
15	0.0642	0.0755	0.0113	0.840	13.45
16	0.0651	0.0904	0.0253	0.930	27.20
17	0.0642	0.0835	0.0193	0.870	22.18
18	0.0650	0.0776	0.0126	0.700	18.00

[1] Before filtering; [2]After filtering. Analyses performed by Environmental Geochemistry Laboratory from UNIR and LABIDRO from IGCE-UNESP.

Table 13. Results of the analyses (in µg/g) of suspended solids from Madeira River between Teotônio and Santo Antônio falls

Sample No.	K	Na	Fe	Cu	Zn	Ba	Cd	Cr	Pb	Ca	Mg	B	Ni	Mn	P
1	31,765	9,412	107,059	< 588	588	941	< 588	< 588	1,176	12,941	11,765	< 588	< 588	1,176	5,882
2	23,809	5,714	61,429	< 238	952	952	< 238	< 238	< 476	8,095	9,524	< 238	< 238	1,428	3,333
3	24,242	6,061	76,364	364	606	5,454	< 303	< 303	606	14,545	9,697	< 303	< 303	1,818	6,667
4	23,415	7,805	75,122	488	488	3,902	< 244	< 244	976	9,756	10,244	< 244	< 244	976	3,415
5	26,846	6,711	80,537	335	671	671	< 335	< 335	671	8,054	10,067	< 335	< 335	671	4,027
6	21,201	5,300	56,184	353	353	706	< 177	< 177	353	6,007	9,187	< 177	< 177	1,060	3,180
7	22,667	6,667	61,333	1,333	889	889	< 222	< 222	889	8,000	9,778	< 222	< 222	889	3,111
8	26,619	7,194	76,978	719	719	719	< 360	< 360	719	12,230	11,511	< 360	< 360	719	4,316
9	26,446	6,611	72,727	413	826	826	< 413	< 413	826	8,264	9,917	< 413	< 413	1,653	4,959
10	27,219	5,917	84,615	592	592	1,183	< 296	< 296	592	10,059	1,.834	< 296	< 296	1,183	4,142
11	26,415	5,660	79,245	755	943	849	< 472	< 472	< 943	8,490	10,375	< 472	< 472	943	9,434
12	26,744	6,395	71,512	581	581	581	< 290	< 290	581	8,721	10,465	< 290	< 290	581	3,488
13	29,677	5,808	87,742	516	645	645	< 322	< 322	< 645	16,774	11,613	< 322	< 322	1,935	4,516
14	27,132	6,977	72,868	775	775	775	< 387	< 387	775	10,853	10,852	<1,550	<387	775	6,356
15	32,743	10,619	81,416	619	885	885	< 442	< 442	885	17,699	10,619	< 442	< 442	1.77	4,425
16	23,715	5,533	74,704	790	395	790	< 198	< 198	395	8,300	10,277	< 198	< 198	1,186	3,162
17	35,751	6,217	64,767	414	518	1,036	< 259	< 259	518	11,917	9,326	< 259	< 259	1,554	4,145
18	27,778	9,524	79,365	794	1,587	794	< 397	< 397	794	16,667	11,111	< 397	< 397	1,587	3,968

Analyses performed by Analytical Chemistry Laboratory from CENA.

reports the results of the chemical analyses of the suspended solids. The average values are: potassium = 26,900 μg/g, sodium = 6,895 μg/g, iron = 75,775 μg/g, copper = 502 μg/g, zinc = 672 μg/g, barium = 827 μg/g, cadmium = 330 μg/g, chromium = 330 μg/g, calcium = 10,965 μg/g, lead = 712 μg/g, magnesium = 10,453 μg/g, boron = 351 μg/g, nickel = 351 μg/g, manganese = 1,216 μg/g and phosphorus = 951 μg/g. The most abundant elements in the particulate matter are: sodium, potassium, iron, calcium and magnesium. The values obtained for silicon ranged from 117 to 353 μg/g and they are also minima as the samples were digested with HF, like in the case of the bottom sediments.

X-Ray Diffraction Analysis

The use of X-ray diffraction has been performed for samples 3 and 18 of bottom sediments. In both cases, it was possible identify the presence of quartz (SiO_2), illite [$K_{0.61}$ $Na_{0.002}$ $Ca_{0.01}$ $Si_{3.42}$ $Al_{0.57}$ $Al_{1.29}$ $Fe_{1.29}$ $Fe_{0.19}$ $Mg_{0.18}$ $Ti_{0.04}$ $O_{10}(OH)_2$], *chlinoclore* (H_8 Mg_5 $Al_2Si_3O_{14}$ = $4H_2$ .5MgO. $Al_2O_3.3SiO_2$), as well mica, feldspars, pyrophyllite, gibbsite, ilmenite, amorphous Fe-Mn oxides, kaolinite, and chlorite-vermiculite mixed layers. Some of these minerals represent alteration products of granitoids, i.e., feldspars yielding kaolinite. Such mineral composition justifies the accentuated concentrations verified for potassium, iron and magnesium. The illite is a non-expansive micaceous clay mineral 2:1 type; it contains an octaedric sheet between two tetraedric ones, constituting one unitary layer of thickness corresponding to about 10 Å (Carrol, 1959), which is linked to the consecutive layer by potassium atoms that neutralize the excess of negative charges in the layers. In most cases, the cation K^+ is not enough to promote the charge balance and the neutralization is held by Na^+ and Ca^{2+} ions present as exchangeable cations (Carrol, 1959). This possibly justifies the high values found for Na and Ca in the bottom sediments and suspended solids.

Statistical Correlation Tests

Statistical tests of correlation were performed considering all data obtained for bottom and suspended sediments. The reference values of the *t* test (statistical distribution of Student), $t_{tab,}$ at 5% significance level and 12 (n = 14) or 16 (n = 18) degrees of freedom, corresponded to 1.782 and 1.746, respectively (Fisher and Yates, 1953). The *t* values were calculated by the equation $t_{calc} = r\ (n\text{-}2/1\text{-}r^2)^{1/2}$ where *r* is the Pearson correlation coefficient. The significance of the correlation was tested using the null hypothesis, which was accepted when $t_{tab} > t_{calc}$. Significant relationships were found between K and Fe (r^2 = 0.64, n = 14), Mg and K (r^2 = 0.83, n = 14), Fe and Mg (r^2 = 0.52, n = 14), and Fe and Mn (r^2 = 0.77, n = 14) for bottom sediments (figure 16), but none significant correlation was identified between K and Na (r = -0.10) and K and Ca (r = 0.39), whose presence has been mainly attributed to illite, as a consequence of the favorable Na and Ca migration to the liquid and suspended solids phases. Significant relationships have also been found between K and Fe (r^2 = 0.23, n =18), Na and Fe (r^2 = 0.24, n =18), Ca and Fe (r^2 = 0.27, n = 18), Mg and Fe (r^2 = 0.62, n = 18) for suspended solids (figure 16), reflecting mainly the composition of Fe-Mn oxides, micas and feldspars in the sediments.

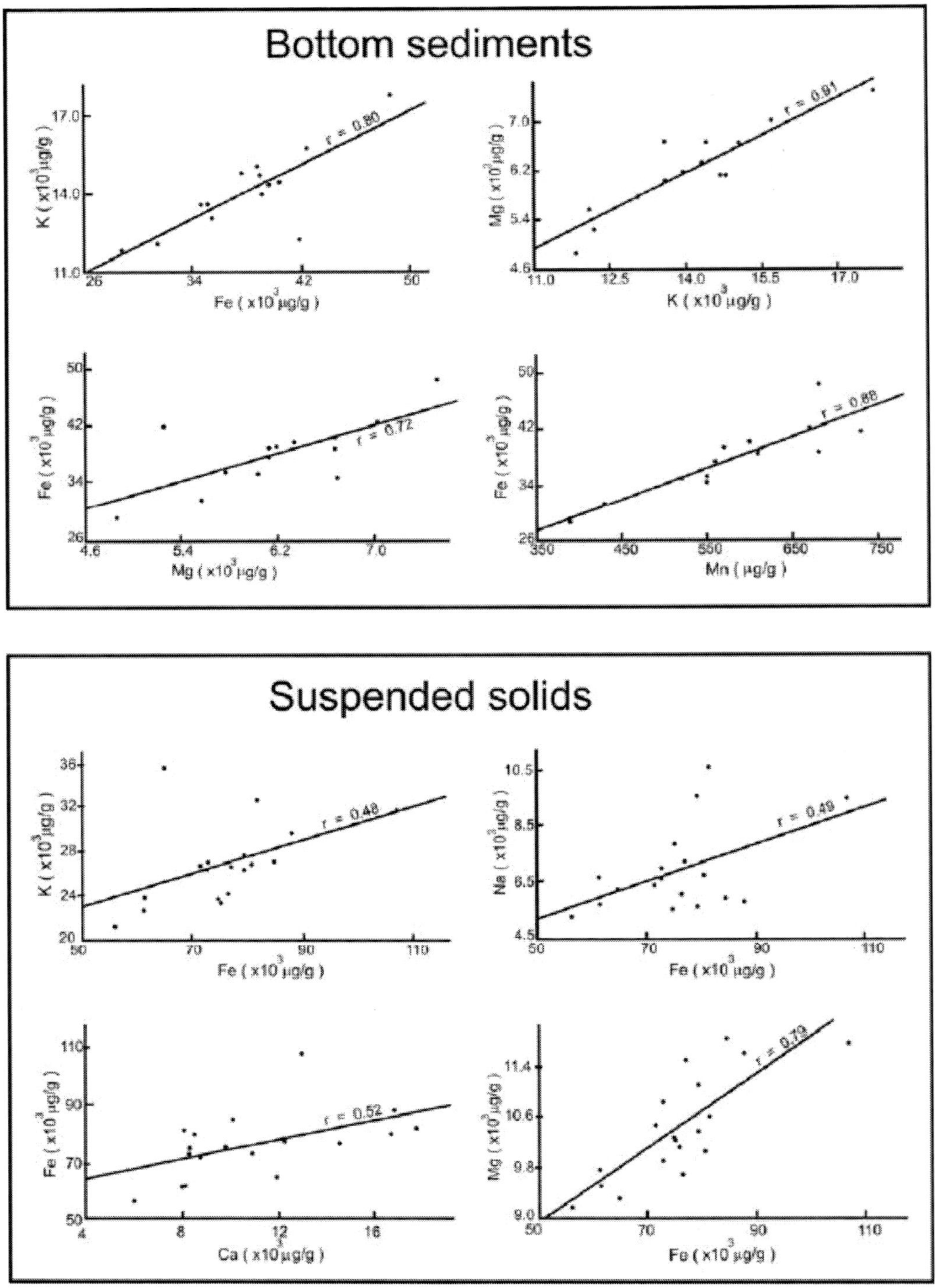

Figure 16. The relationships between potassium, sodium, calcium, magnesium, iron, and manganese in bottom sediments and suspended solids from Madeira River between Teotônio and Santo Antônio falls.

No significant correlation was found between the Hg content and other chemical elements analyzed in bottom sediments (table 14), what is surprising since, among other reasons, Fe, Al and Mn oxy-hydroxides have been recognized to have an important role in retaining/transporting the Hg in drainage systems of the Amazon basin (Rodrigues Filho and Maddock, 1997; Roulet et al., 2000). The grain size of the sediments was not evaluated because the size fraction utilized for chemical analysis (<74 μm) is close to the fraction <63 μm often used in numerous metal studies, allowing a comparison of results (Förstner and Salomons, 1980). Thus, a possible explanation for the lack of correlation with Hg in the studied reach of the Madeira River consists on the small size (area of about 260 km^2, along 25 km) of the sector utilized for sampling that did not exhibit a great variability in the chemical analyses. For instance, if it is considered the larger area (about 6×10^4 km^2, along 700 km) investigated by Vergotti (2002), that included eight lakes in this drainage basin, it is possible to identify significant correlations between Hg, organic matter, and R_2O_3 (index proposed by Colman (1982) to group Al_2O_3, Fe_2O_3, and TiO_2), suggesting the Hg adsorption by organic matter (Loring, 1975), kaolinite, and Fe and Al oxy-hydroxides.

Table 14. Correlation coefficients among mercury and others elements/compounds analyzed for bottom sediments from Madeira River between Teotônio and Santo Antônio falls

ELEMENT/COMPOUND	CORRELATION COEFFICIENT WITH MERCURY
Potassium	0.07
Sodium	0.29
Iron	0.08
Copper	- 0.13
Zinc	-0.24
Barium	-0.03
Cadmium	0.12
Chromium	0.23
Lead	0.07
Calcium	0.27
Magnesium	0.03
Boron	-0.14
Nickel	-0.5
Manganese	0.36
Organic Matter	-0.09
Phosphorus	0.02

Chapter 8

THE MOBILITY OF MERCURY AND OTHER ELEMENTS IN THE MADEIRA RIVER

The concentration data for mercury and others elements in waters, sediments and suspended solids from Madeira River between Teotônio and Santo Antônio falls allowed estimate mobility coefficients that were useful for providing a comparative evaluation of their behavior there. This chapter describes the major advances attained from the acquired data.

THE PREFERENCE RATIO RELATIVE TO MERCURY

In order to express how minor elements may enter crystals forming from a cooling melt or from an aqueous solution, the geochemists involved with petrology defined the distribution coefficient as the ratio between the concentration of the minor element in the crystal of a mineral and the concentration of that element in the fluid from which the crystal is formed (Faure, 1991; Richardson and McSween, 1989; Krauskopf and Bird, 1995). In some cases, when it has been necessary to measure the partitioning of a particular chemical species between sediments and water, the distribution coefficient, K_D, has been expressed by the relation (Borovec, 1981; Benes, 1984): $K_D = (m_a/m_s) \times V/m$ where V/m is the liquid/solid ratio (V is the volume of water and m is the weight of sediments), m_a and m_s are amounts (weight) of the chemical species adsorbed and remaining in solution, respectively. Such a parameter has also been named as geochemical enrichment factor (Szalay, 1964), enrichment factor (Langmuir, 1978), or adsorption coefficient (AAEC, 1983), being expressed in cm^3g^{-1} or m^3kg^{-1}.

Another not dimensionless K_D definition involving the affinity of mercury for the particulate phase can be written by the equation: $K_D = C_p / C_d$ where C_p is the particulate Hg content (in ng kg^{-1}) and C_d is the dissolved Hg concentration (in ng L^{-1}). According to these units, K_D is expressed in $L.kg^{-1}$, with higher K_D´s suggesting a stronger Hg affinity by the particulate fraction.

However, some authors also preferred to express K_D as a dimensionless parameter, considering a solution of diluted water (density = 1 g cm^{-3} = 10^3 kg m^{-3}) and the assumption that 1 cm^3 is equal to 1 g (or 1 m^3 is equal to 10^3 kg). For instance, some dimensionless values reported by AAEC (1983) for natural materials with respect to UO_2^{2+} are: $1.1\text{-}2.7\times10^6$

for amorphous Fe(III) oxyhydroxides, 10^4-10^6 for peat, 4×10^3 for fine-grained goethite, 10^3 for montmorillonite and 50 for kaolinite.

Despite these K_D's definitions, with the purpose of evaluating the preference for Hg relative to other elements considered in this investigation, two dimensionless ratios will be utilized:

(1) $P_{l.phase}$ = $(Hg/element)_{liquid\ phase}$ / $(Hg/element)_{bottom\ sediments}$ which expresses the ratio involving the Hg and other elements in the liquid phase divided by the ratio involving the Hg and other elements in bottom sediments.

(2) $P_{s.solids}$ = $(Hg/element)_{suspended\ solids}$ / $(Hg/element)_{bottom\ sediments}$ which is defined as the ratio involving the Hg and other elements in the particulate matter divided by the ratio involving the Hg and other elements in bottom sediments.

These preference ratios are useful to compare the Hg transport among three different phases: liquid, bottom sediments and particulate matter. They also are of easy calculation, since the same results are generated for $P_{l.phase}$ and $P_{s.solids}$ if the concentrations obtained from the chemical data are utilized instead of the molar ratio Hg/element, having already been successfully applied to the evaluation of the uranium migration to waters interacting with carbonates and granites (Bonotto and Andrews, 1993, 2000; Bonotto et al., 2001).

THE PREFERENCE FOR HG IN MADEIRA RIVER

In order to utilize the Hg preference ratios in Madeira River between Teotônio and Santo Antônio falls, it was performed the calculations considering the average concentrations reported in tables 8-13, and the detection limit of the ICP-AES method in the liquid phase for the following elements: Mn = 0.02 mgL^{-1}, Cu = 0.05 mgL^{-1}, Zn = 0.05 mgL^{-1}, Pb = 0.1 mgL^{-1}, Ni = 0.05 mgL^{-1}, Cr = 0.05 mgL^{-1}, Cd = 0.05 mgL^{-1}, Ba = 0.05 mgL^{-1}, B = 0.05 mgL^{-1}, and P = 0.1 mgL^{-1}.

Three different evaluations were realized, i.e. using the total mean value of the Hg content in the liquid phase (13.8 ngL^{-1}) and particulate matter (10.8 ngL^{-1}) determined by Nriagu et al. (1992) (table 9), the mean value of the Hg content in the liquid phase (14.4ngL^{-1}) and particulate matter (9.3 ngL^{-1}) determined by Nriagu et al. (1992) in the same sampling points of this investigation (table 9), and the upper Hg concentration of 0.5 μgL^{-1} in surface and ground waters established by WHO (1993). Table 15 reports the results obtained in the calculations.

Initially, since the average Hg concentrations measured by Niagru (1992) in the liquid phase (13.8 and 14.4 ngL^{-1}) and particulate matter (10.8 and 9.3 ngL^{-1}) from Madeira River at and distant of the studied area do not differ significantly, it is possible to verify that the Hg preference ratios are practically the same (table 15).

The values reported in table 15 also indicate that Hg is never mobilized preferentially to all other elements in suspended solids, since the $P_{s.solids}$ ratios are greatly lower than one. Mercury only could be transported preferentially to all other analyzed elements in the particulate phase under two non-realistic situations, i.e. if its concentration would reach values at least 10^4-fold higher than those expected/quantified in the area, or, alternatively, considering the same Hg concentration, if the analyzed elements would reach values 10^4-fold lower than those actually quantified. Even rainstorms are expected not rise up to 10^4-fold higher the Hg concentration measured during dry conditions (Lacerda et al., 1995), implying that the lack of the hydrological period in table 9 is not critical for the calculations affecting the particulate matter.

Table 15. Preference ratios for Hg relative to other elements in Madeira River between Teotônio e Santo Antônio falls

ELEMENT	$P_{l.phase}$ [1]	$P_{s.solids}$ [2] ($\times 10^{-4}$)	$P_{l.phase}$ [3]	$P_{s.solids}$ [4] ($\times 10^{-4}$)	$P_{l.phase}$ [5]
Na	0.47	1.49	0.49	1.28	16.96
K	2.50	1.18	2.60	1.01	90.44
Ca	0.09	0.62	0.09	0.54	3.14
Mg	0.46	1.34	0.48	1.15	16.83
Fe	84.26	1.13	87.92	0.97	3052.96
Mn	8.37	1.08	8.73	0.93	303.12
Cu	3.66	2.85	3.82	2.45	132.50
Zn	2.42	1.41	2.53	1.21	87.71
Pb	0.48	0.53	0.51	0.46	17.60
Cr	0.03	0.03	0.03	0.03	1.04
Cd	0.03	0.03	0.03	0.03	1.04
Ni	0.31	0.35	0.32	0.30	11.25
Ba	0.59	0.28	0.62	0.24	21.46
B	0.07	0.08	0.08	0.07	2.71
P	2.73	0.48	2.85	0.41	99.06
Range	0.03-84.26	0.03-2.85	0.03-87.92	0.03-2.45	1.04-3052.96
Max/Min ratio	2930.85	83.58	2930.85	83.33	2930.84

[1] $P_{l.phase} = (Hg/element)_{liquid\ phase}/(Hg/element)_{bottom\ sediments}$; $(Hg)_{liquid\ phase} = 13.8$ ngL^{-1} (Nriagu, et al.,1992)

[2] $P_{s.solids} = (Hg/element)_{suspended\ solids}/(Hg/element)_{bottom\ sediments}$; $(Hg)_{suspended\ solids} = 10.8$ ngL^{-1} (Nriagu, et al., 1992).

[3] $P_{l.phase} = (Hg/element)_{liquid\ phase}/(Hg/element)_{bottom\ sediments}$; $(Hg)_{liquid\ phase} = 14.4$ ngL^{-1} (Nriagu, et al., 1992).

[4] $P_{s.solids} = (Hg/element)_{suspended\ solids}/(Hg/element)_{bottom\ sediments}$; $(Hg)_{suspended\ solids} = 9.3$ ngL^{-1} (Nriagu, et al., 1992).

[5] $P_{l.phase} = (Hg/element)_{liquid\ phase}/(Hg/element)_{bottom\ sediments}$; $(Hg)_{liquid\ phase} = 0.5$ μgL^{-1} (WHO, 1993).

The suggestion that Hg concentration in water is independent of Hg content in sediments, remaining constant during rainstorms (Lacerda et al., 1995), also implies that the lack of the hydrological period in table 9 is not relevant for the calculations involving the liquid phase. However, the situation is more complex when the liquid phase is considered in relation to the bottom sediments, since it was possible to find values indicating that Hg is mobilized less preferentially than Na, Ca, Mg, Pb, Cr, Cd, Ni, Ba, and B ($P_{l.phase}$ ratios lower than unity), as well as that Hg is transported more preferentially than K, Fe, Mn, Cu, Zn, and P ($P_{l.phase}$ ratios higher than one) (table 15). An additional difficulty on the interpretation resides in the fact that if it is considered the calculation taking into account the upper Hg concentration of 0.5 μgL^{-1} in surface and ground waters established by WHO (1993), that is 36-fold higher than the average value reported in table 9, then, it could be suggested that Hg always is mobilized preferentially to all other elements in the liquid phase, since the generated $P_{l.phase}$ ratios are greater than one (table 15). Thus, these calculations indicate that it is not necessary a 10^4-fold increment on the Hg content in the liquid phase to drastically modify the interpretation of its mobility relative to that of others elements.

Another important aspect that should be discussed involving the liquid phase is related to the use on the calculations of a very low Hg content (13.8 ngL^{-1}) and of the detection limit of the ICP-AES method instead of actually measured concentrations for the elements Mn, Cu, Zn, Pb, Ni, Cr, Cd, Ba, B, and P. If this aspect is avoided, then, a more appropriate evaluation should restrict to Fe and major cations Na, K, Ca, and Mg. Among them, the generated $P_{l.phase}$ ratios are greater than one for K and Fe (table 15), suggesting, initially, that these two very common elements may be less mobile than Hg in the liquid phase at the studied part of Madeira River between Teotônio e Santo Antônio falls. However, another possibility that cannot be disregarded is a lower presence of these elements in the system, due to a strong removal from rocks, soils and recent (Cenozoic) sediments, which is related to the climate of the area, i.e. average annual temperature of 25.5°C, average annual humidity of 83.3%, and annual precipitation of 1625 mm (Lisboa, 1989; FIBGE, 1991).

Isovolumetric geochemical balances have been performed by Silva (1995) for several weathered profiles from the Brazilian Amazon, with the enrichment factor (EF) for a constituent being calculated by the equation: $EF = E_w D_{rm} / E_{rm} D_w$ where E_w, E_{rm} is the concentration of the constituent (in wt %) in the weathered material and rock matrix (original material), respectively, and D_w, D_{rm} is the density (in gcm^{-3}) of the weathered material and rock matrix (original material), respectively. Values lower than one have been obtained for EF relative to K_2O and Fe_2O_3 in alluvial soils developed over recent alluvial deposits, which typically represent the main materials leached by the waters of the Madeira River. Thus, such EF´s values lower than one suggest the removal of K and Fe from the system, reinforcing their little presence in the studied area, as well indicated the calculated $P_{l.phase}$ ratios greater than one (table 15).

The Preference Ratios for Other Elements in Madeira River

The usefulness of these dimensionless preference ratios for any analyzed element in Madeira River between Teotônio e Santo Antônio falls was also evaluated, taking into account the average concentrations reported in tables 8-13 (except for mercury that was already considered) and the detection limit of the ICP-AES method in the liquid phase for the elements Mn = 0.02 mgL^{-1}, Cu = 0.05 mgL^{-1}, Zn = 0.05 mgL^{-1}, Pb = 0.1 mgL^{-1}, Ni = 0.05 mgL^{-1}, Cr = 0.05 mgL^{-1}, Cd = 0.05 mgL^{-1}, Ba= 0.05 mgL^{-1}, B = 0.05 mgL^{-1}, and P = 0.1 mgL^{-1}. Table 16 reports the minimum and maximum values obtained in the calculations, as well the maximum/minimum (Max/Min) ratio quantified from the observed range.

The most prominent aspect to be observed from the data set in table 16 is that the $P_{l.phase}$ and $P_{s.solids}$ ratios do not exhibit a very narrow range for any particular element. However, when the maximum value is divided by the minimum one, then, a surprising almost constant value is found for all elements, i.e. 2931 for $P_{l.phase}$ and 84 for $P_{s.solids}$. The same constant values were also calculated for the three different evaluations involving mercury (table 15), implying, firstly, that the preference ratios may be useful predictors for larger systems including elements with variable chemical properties/ geochemical behavior, and, secondly, that the use of older Hg data do not affect the Max/Min preference ratios. Such finding as well suggests that similar Max/Min preference ratios would be obtained if more precise values lower than the detection limit of the ICP-AES method in the liquid phase for the elements Mn, Cu, Zn, Pb, Ni, Cr, Cd, Ba, B, and P were utilized in the calculations.

Fe oxides and hydroxides are highly sorptive materials, being extensively recognized to have an important role in retaining elements in aquatic systems [for instance, see Langmuir (1978), AAEC (1983), Rodrigues Filho and Maddock (1997), among others]. Such aspect was clearly evidenced by their presence in bottom sediments at the studied area, and because even the maximum iron $P_{l.phase}$ reported in table 16 is greatly lower than one, differently from the remnant maximum calculated values higher than one. The sorption property of Fe oxides and hydroxides implies on a Max/Min preference ratio corresponding to 291 (table 16), that is 10-fold lower than other calculated values, a variability not much pronounced when compared with those conventionally generated by K_D formula, but extremely important to indicate the sensitivity of $P_{l.phase}$ to the dominance of sorption processes. In terms of $P_{s.solids}$, the Max/Min preference ratio corresponding to 44 for Cu and Ni (table 16) only was 2-fold lower than other calculated values.

Table 16. Preference ratios for samples collected at Madeira River between Teotônio and Santo Antônio falls

ELEMENT	$P_{l.phase}$ [1]			$P_{s.solids}$ [2]		
	Minimum	Maximum	Max/Min ratio	Minimum	Maximum	Max/Min ratio
Na	0.061	180.04	2930.87	0.023	1.92	83.62
K	0.012	33.76	2930.83	0.029	2.42	83.62
Ca	0.33	972.75	2930.84	0.055	4.59	83.62
Mg	0.062	181.38	2930.83	0.026	2.13	83.62
Fe	0.00034	0.099	291.17	0.030	2.52	83.62
Mn	0.0034	10.07	2931.21	0.032	2.65	83.62
Cu	0.0079	23.04	2930.85	0.012	0.52	43.63
Zn	0.012	34.81	2930.97	0.024	2.02	83.62
Pb	0.059	173.42	2930.82	0.064	5.34	83.62
Cr	1.00	2930.85	2930.85	1.00	83.62	83.62
Cd	1.00	2930.85	2930.85	1.00	83.62	83.62
Ni	0.092	271.37	2930.83	0.098	4.30	43.63
Ba	0.048	142.27	2930.83	0.12	10.17	83.62
B	0.38	1127.25	2930.85	0.41	34.21	83.62
P	0.010	30.82	2930.92	0.071	5.97	83.62

[1] $P_{l.phase} = (\text{Element} / \text{Element}_j)_{\text{liquid phase}} / (\text{Element} / \text{Element}_j)_{\text{bottom sediments}}$; $j = 1, 2, \ldots n$

[2] $P_{s.solids} = (\text{Element} / \text{Element}_j)_{\text{suspended solids}} / (\text{Element} / \text{Element}_j)_{\text{bottom sediments}}$; $j = 1, 2, \ldots n$

Chapter 9

MERCURY IN LAKE SEDIMENTS FROM MADEIRA RIVER BASIN

In aquatic systems, heavy metals (including mercury) and other elements are mainly transported in the solid phase, either sorbed onto particle surfaces and coatings, or incorporated into mineral grains. Lakes located at the Madeira River basin have been affected by mercury inputs, mainly during the peak activities related to the gold mining. This chapter describes the results obtained in the chemical analyses of sediments collected from nine lakes located at the Madeira River basin, Rondônia State, with the objective of investigating the influence of organic matter on the mercury retention.

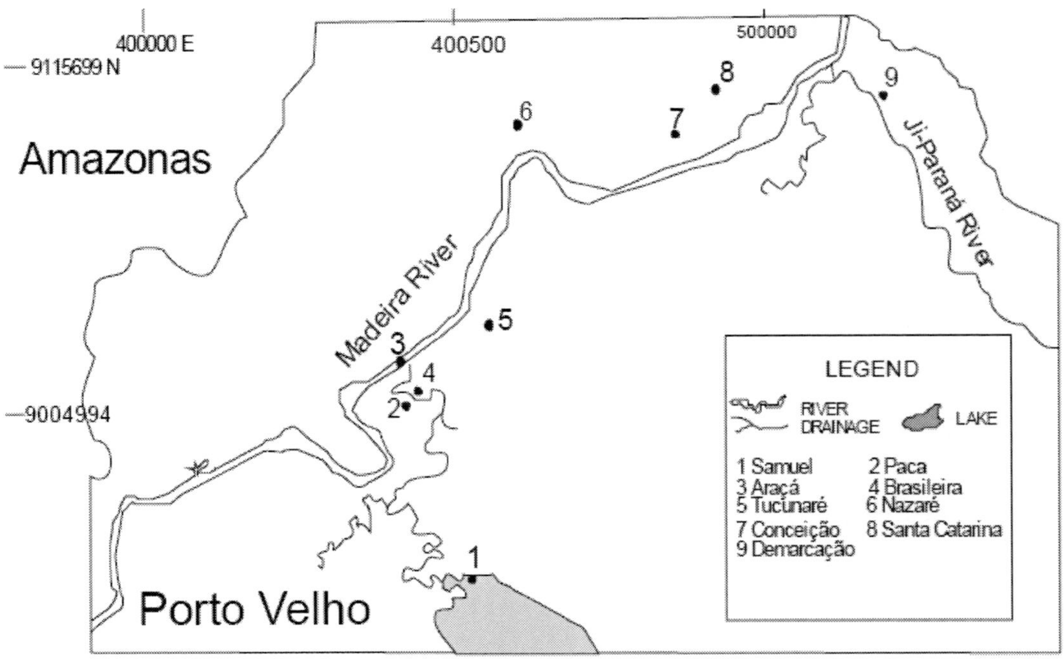

Figure 17. Location of the lake sediment sampling sites for Hg analysis at the Madeira River basin, Rondônia State, Brazil. According to Bonotto et al. (2006).

SAMPLING

Core sediments (20 to 80 cm deep) were collected at nine lakes between the Jamari and Ji-Paraná rivers that are tributaries of Madeira River; this area is situated between 8° and 9°S and between 62° and 64°W (figure 17). The fieldwork was undertaken during October 1999 and February 2000 after utilizing a small boat that crossed several *igarapés* for reaching the lakes. The lakes exhibited variable dimensions (5 to 10 m deep, 600 to 7000 m long, and 300 to 1200 m wide), with clear and black waters, depending on the dissolved organic matter content. A 1 m long acrylic tube (10 cm in diameter) adapted to an iron outliner was driven into the sediments preferentially at the middle of each lake, each profile being cut in 5 cm thick slices that had been transferred to polyethylene bags. The major features of the lakes are:

1. Samuel: it comprises a dam to generate electricity, whose activities started in 1989. The surface area of the reservoir is about 450 km^2, one of the largest in the Amazon area. It is situated in the Jamari River channel, and the access may be realized by BR-364 road in the Porto Velho-Cuiabá direction. The waters of this lake are clear and the sediments sampled are yellow and slightly sandy.
2. Paca: it is situated at the opposite (right) margin of the channel that supplies water to Brasileira lake, close to the confluence of Jamari River to Madeira River. It is directly influenced by Jamari River, practically consisting on a small flood plain area. The waters are clear and the size a little lower than Tucunaré lake (diameter ~ 600 m).
3. Araçá: it is situated at the left margin of Jamari River, about 500 m from the confluence with Madeira River. The waters are clear and the sediments are dark clayey. The approximate dimensions are 500 × 2000 m and the channel width is about 5 m, which allows the discharge to Jamari River. The lake becomes isolated during the dry period.
4. Brasileira: it is located 5 km upstream from the Jamari River mouth, where São Carlos district is situated. The approximate dimensions are 2000 × 500 m, the waters are clear and it receives water from a 10 m wide channel that links it directly to Jamari River. The samples collected are dominantly composed by clear clay, without the sand presence, also occurring many leaves and others organic materials.
5. Tucunaré: it is situated 10 km from Jamari River mouth at the right margin of Madeira River, more specifically at *Terra Caída* village, that is situated at the opposite (left) margin. The access is obtained from Tucunaré *igarapé*. The sediments sampling was realized at the peak of the dry season, when the waters were retained (approximate dimension of 800 × 300 m). The waters of this lake are clear and the sediments are clayey.
6. Nazaré: it is situated at Nazaré village (left margin of Madeira River), and the access is obtained from Nazaré *igarapé*. The waters are black, as well the sediments, which also contain a great amount of leaves. It represents a large area of flooded forest and the water speed is low. The flooded forest areas are situated about 100 m between themselves and at the dry season they should correspond to their margins.

7. Conceição: its access is obtained from Conceição *igarapé*, and the characteristics are similar to those of Santa Catarina lake in terms of the waters and sediments colors. The approximate dimensions are 1200 × 5000 m.
8. Santa Catarina: it is situated at the left margin of Madeira River. The access is obtained from Conceição *igarapé* that discharges at Madeira River at the Conceição village. The waters and sediments are black and it is the largest of the lakes investigated (the approximate dimensions are 7000 × 1000 m).
9. Demarcação: the Demarcação village is located at the right margin of Machado River, about 10 km from its mouth at Calama district. It is situated upstream from Demarcação village, at the right margin of Machado River, and the access occurs by the first *igarapé*. The waters are black and their speed is low. The sampled sediments are black and they contain an accentuated presence of detritic materials. The margins of this lake were flooded during sampling, thus, it had expanded inside the forest. The approximate radius of the lake is 200 m.

Analytical Protocol

All bags containing the sediments were stored in iceboxes and transported to the laboratory. The material was dried at 60°C, crushed, quartered, sieved for separation of the < 200 mesh (< 74 μm) size fraction, and divided into different aliquots for analysis of the total Hg content, organic matter, loss on ignition (LOI), and major oxides.

Digestion of the samples and Hg extraction steps were used, as described by Malm et al. (1989). About 0.5 g of each sample was used, 1 mL of Milli Q water plus 2 mL of aqua regia were added, and heated to 50°C for 5 min. After cooling, 10 mL of Milli Q water plus 4 mL of 5% potassium permanganate were added and heated to 55°C for 30 min. The excess $KMnO_4$ was neutralized with 12% hydroxylamine hydrochlorate. Each aliquot representing the sample was divided in three sub-aliquots. The analyses of total Hg concentrations were carried out using atomic absorption spectrophotometry with cold vapour generation. The equipment is a Perkin Elmer model FIMS-400 atomic absorption spectrophotometer installed at the Environmental Geochemistry in UNIR, Porto Velho city.

The organic matter content was evaluated by spectrophotometry (Hach, 1992) held in a solution obtained after adding potassium dichromate and sulphuric acid to 1 g powdered aliquot not submitted to digestion; organic carbon was oxidized to carbon dioxide with a parallel reduction of hexavalent chromium to trivalent chromium and an accompanying colour change from orange to green read at 610 nm (Hach, 1992).

Another (indirect) evaluation of the organic matter content in the lake sediments samples was determined by measuring weight loss in sub-samples after burning at selected temperatures. Since the samples are from freshwater, the majority of "loss-on-ignition" (LOI) refers to organic carbon oxidizing to CO_2. In general, these results are typically accurate to 1-2% for sediment with over 10% organic matter, where, in high clay sediment, water of hydration may be lost during the ignition, resulting in additional error. Additional steps to subtract out weight loss due to the oxidation of sulfur to SO_2 were not used, as this procedure is often applied to samples from saline environments. The procedure utilized was to weigh the crucible using the mass balance; place a few grams of moist sub-sample in the crucible and measure the combined mass; dry the sample in the oven at 100°C; re-weigh the crucible + sub-sample (the difference yields the water content); set the muffle furnace temperature to 450°C and leave the crucible in it for 8 hours; remove crucible from furnace and allow to cool briefly; re-weigh crucible + sub-sample (the difference from the dry state yields the organic content).

The chemical characterization of the major oxides (SiO_2, Al_2O_3, Na_2O, K_2O, CaO, MgO, MnO, Fe_2O_3, TiO_2 and P_2O_5) was realized by X-ray fluorescence (equipment Philips, model PW2510) at the Geochemistry Laboratory from UNESP-University of the State of São Paulo, Rio Claro.

RESULTS OF THE CHEMICAL ANALYSIS

All chemical data for Hg, organic matter, LOI and major oxides contents in sediments lakes from Madeira River basin are given in tables 17-30, as reported by Vergotti (2002). Table 31 reports the average composition of the lake sediments, also including the data for total bases ($MgO+CaO+Na_2O+K_2O$) and oxides resistant to weathering, i.e. $Al_2O_3+Fe_2O_3+TiO_2$ (R_2O_3), according to Colman (1982).

Major Significant Relationships

The average Hg content in the lake sediments from Madeira River basin ranged between 33 and 157 ng/g, which is similar to the range of 21-136 ng/g determined for the bottom sediments between Teotônio and Santo Antônio waterfalls (table 10).

Table 17. Hg content (in ng/g) of lake sediments from the Madeira River basin, Rondônia State, Brazil. According to Vergotti (2002)

Depth (cm)	Samuel	Paca	Araçá	Brasileira	Tucunaré	Nazaré	Conceição	Santa Catarina	Demarcação
0-5	126.2	128.8	58.5	76.5	48.8	33.8	56.6	49.7	164.9
5-10	162.9	145.3	55.2	78.8	44.7	34.0	48.4	57.6	110.3
10-15	171.4	148.4	46.8	113.8	45.4	23.7	47.3	82.4	92.9
15-20	168.2	58.2	42.6	120.0	39.6	30.9	50.8	83.4	84.2
20-25		175.5	42.5	118.6	34.0	41.3	49.6	74.2	99.6
25-30			47.4	115.4	39.3		80.6	74.3	100.3
30-35			44.2	114.6	36.5		62.4	72.4	97.8
35-40			37.2	127.0	36.7		66.9	82.2	
40-45			34.0	153.2	38.1		66.1	77.0	
45-50			33.8	131.7	49.0		65.6	95.0	
50-55			35.8	119.6	38.1				
55-60			36.5		39.0				
60-65			33.8		40.2				
65-70			39.7						
70-75			50.2						
75-80			49.5						

Table 18. The organic matter content (in %) of lake sediments from the Madeira River basin, Rondônia State, Brazil. According to Vergotti (2002)

Depth (cm)	Samuel	Paca	Araçá	Brasileira	Tucunaré	Nazaré	Conceição	Santa Catarina	Demarcação
0-5	17.8	9.7	4.0	6.0	3.9	3.7	9.1	5.0	10.1
5-10	18.0	8.6	3.1	4.9	3.5	1.9	6.1	8.6	6.5
10-15	18.7	9.8	7.1	5.6	3.8	3.8	7.1	3.9	6.1
15-20	17.8	9.3	6.9	4.0	5.5	3.9	8.0	5.0	6.9
20-25		9.4	4.0	8.2	3.6	4.0	6.7	6.8	6.9
25-30			7.1	4.8	4.0		7.0	12.8	7.8
30-35			5.0	5.0	3.0		7.8	5.9	8.8
35-40			4.8	4.0	2.0		7.7	4.6	
40-45			20.5	4.7	3.6		8.8	5.7	
45-50			2.0	3.8	3.0		9.1	2.7	
50-55			4.0	4.0	3.6				
55-60			2.0		2.8				
60-65			5.0		4.6				
65-70			7.1						
70-75			4.2						
75-80			4.2						

Table 19. The LOI (loss on ignition) content (in %) of lake sediments from the Madeira River basin, Rondônia State, Brazil. According to Vergotti (2002)

Depth (cm)	Samuel	Paca	Araçá	Brasileira	Tucunaré	Nazaré	Conceição	Santa Catarina	Demarcação
0-5	16.1	13.9	9.6	10.5	6.8	6.9	11.6	8.0	14.4
5-10	15.1	14.0	9.0	10.5	7.0	6.9	9.9	7.6	11.7
10-15	14.8	14.7	8.5	12.3	6.6	6.6	11.3	7.5	11.0
15-20	14.5	7.4	8.2	11.1	6.1	6.8	10.6	7.2	11.1
20-25		16.0	7.8	11.0	6.1	7.1	10.8	7.0	11.3
25-30			7.9	11.7	6.5		11.1	7.3	11.3
30-35			7.7	11.6	5.8		11.4	7.4	11.5
35-40			6.6	11.6	6.2		12.0	7.5	
40-45			6.3	11.7	6.2		12.2	7.2	
45-50			6.6	11.8	6.0		12.9	7.3	
50-55			6.7	11.5	6.4				
55-60			7.5		6.3				
60-65			7.7		5.6				
65-70			7.7						
70-75			7.4						
75-80			7.2						

Table 20. The moisture content (in %) of lake sediments from the Madeira River basin, Rondônia State, Brazil. According to Vergotti (2002)

Depth (cm)	Samuel	Paca	Araçá	Brasileira	Tucunaré	Nazaré	Conceição	Santa Catarina	Demarcação
0-5	40.5	-	54.0	54.0	47.2	42.3	63.9	50.1	42.9
5-10	27.4	-	49.4	44.2	42.8	38.7	51.6	49.9	38.9
10-15	24.3	-	45.9	43.9	49.0	38.8	63.2	47.6	47.9
15-20	21.7	-	54.7	38.8	41.8	36.5	29.1	41.2	48.9
20-25		-	56.2	36.8	44.4	40.8	53.9	44.3	44.1
25-30			46.1	40.4	38.3		72.8	47.0	46.3
30-35			46.5	34.7	33.2		54.3	41.6	45.5
35-40			37.4	34.5	39.4		58.2	45.9	
40-45			39.8	36.3	39.4		55.5	45.3	
45-50			33.1	36.7	40.9		53.4	44.9	
50-55			34.5	33.0	58.2				
55-60			37.1		69.4				
60-65			42.2		34.7				
65-70			41.2						
70-75			43.3						
75-80			44.8						

Table 21. The SiO_2 content (in %) of lake sediments from the Madeira River basin, Rondônia State, Brazil. According to Vergotti (2002)

Depth (cm)	Samuel	Paca	Araçá	Brasileira	Tucunaré	Nazaré	Conceição	Santa Catarina	Demarcação
0-5	43.74	52.44	53.54	55.08	61.54	60.75	55.64	57.89	51.16
5-10	42.55	52.25	55.04	53.82	61.65	60.84	55.41	58.30	53.47
10-15	42.61	52.16	56.36	51.92	60.80	61.18	55.69	59.23	53.69
15-20	48.25	61.71	56.81	52.83	62.79	60.26	55.91	59.68	54.19
20-25		50.20	58.67	53.48	61.94	58.73	56.71	60.35	53.50
25-30			57.97	52.45	60.75		58.24	59.88	53.37
30-35			57.97	52.06	63.78		58.79	59.79	53.00
35-40			59.33	52.06	62.96		58.47	59.85	
40-45			60.99	52.00	62.28		56.73	60.60	
45-50			60.61	52.84	61.90		54.21	60.32	
50-55			60.63	52.32	62.82				
55-60			59.12		62.93				
60-65			58.98		64.74				
65-70			58.39						
70-75			62.20						
75-80			62.64						

Table 22. The Al_2O_3 content (in %) of lake sediments from the Madeira River basin, Rondônia State, Brazil. According to Vergotti (2002)

Depth (cm)	Samuel	Paca	Araçá	Brasileira	Tucunaré	Nazaré	Conceição	Santa Catarina	Demarcação
0-5	29.40	24.05	22.23	22.78	18.39	19.79	22.22	21.41	23.12
5-10	31.43	24.29	21.51	25.59	18.43	19.72	22.78	21.54	22.82
10-15	31.26	24.51	21.27	27.86	19.36	19.65	22.31	21.15	22.54
15-20	27.81	19.62	21.46	27.87	18.31	20.26	22.73	21.12	22.19
20-25		26.86	20.70	27.39	18.79	21.01	22.36	20.84	22.36
25-30			20.95	28.15	19.44		21.19	21.08	22.35
30-35			20.89	28.84	17.84		20.82	21.13	22.68
35-40			20.20	29.03	18.23		21.33	20.79	
40-45			19.27	29.34	18.63		22.68	20.47	
45-50			19.31	28.68	19.01		24.75	20.50	
50-55			19.43	29.41	18.27				
55-60			19.80		18.17				
60-65			19.82		17.31				
65-70			20.71						
70-75			19.37						
75-80			18.92						

Table 23. The Na_2O content (in %) of lake sediments from the Madeira River basin, Rondônia State, Brazil. According to Vergotti (2002)

Depth (cm)	Samuel	Paca	Araçá	Brasileira	Tucunaré	Nazaré	Conceição	Santa Catarina	Demarcação
0-5	0.07	0.33	0.66	0.40	0.82	0.74	0.48	0.67	0.21
5-10	0.08	0.34	0.65	0.41	0.75	0.77	0.57	0.70	0.22
10-15	0.09	0.29	0.68	0.20	0.78	0.76	0.50	0.69	0.24
15-20	0.08	0.76	0.67	0.23	0.81	0.75	0.52	0.71	0.25
20-25		0.15	0.71	0.19	0.84	0.70	0.49	0.70	0.26
25-30			0.69	0.19	0.80		0.45	0.70	0.23
30-35			0.69	0.18	0.84		0.40	0.69	0.23
35-40			0.77	0.15	0.81		0.37	0.69	
40-45			0.80	0.14	0.84		0.34	0.71	
45-50			0.78	0.16	0.81		0.33	0.70	
50-55			0.77	0.18	0.81				
55-60			0.73		0.85				
60-65			0.77		0.84				
65-70			0.71						
70-75			0.75						
75-80			0.78						

Table 24. The K_2O content (in %) of lake sediments from the Madeira River basin, Rondônia State, Brazil. According to Vergotti (2002)

Depth (cm)	Samuel	Paca	Araçá	Brasileira	Tucunaré	Nazaré	Conceição	Santa Catarina	Demarcação
0-5	0.03	2.20	3.23	2.30	2.84	3.26	3.05	3.42	1.88
5-10	0.05	2.17	3.24	2.36	2.89	3.20	3.39	3.43	2.21
10-15	0.03	2.09	3.24	1.81	3.08	3.18	3.13	3.40	2.21
15-20	0.03	2.96	3.35	2.04	2.96	3.25	3.33	3.34	2.16
20-25		1.72	3.30	2.13	3.10	3.31	3.08	3.37	2.19
25-30			3.20	1.91	3.01		2.77	3.35	2.12
30-35			3.25	1.91	2.98		2.53	3.39	2.14
35-40			3.37	1.85	3.03		2.19	3.36	
40-45			3.26	1.82	3.11		2.25	3.30	
45-50			3.32	1.92	3.16		2.40	3.26	
50-55			3.26	1.88	2.96				
55-60			3.23		2.94				
60-65			3.31		2.86				
65-70			3.33						
70-75			2.89						
75-80			2.92						

Table 25. The CaO content (in %) of lake sediments from the Madeira River basin, Rondônia State, Brazil. According to Vergotti (2002)

Depth (cm)	Samuel	Paca	Araçá	Brasileira	Tucunaré	Nazaré	Conceição	Santa Catarina	Demarcação
0-5	0.04	0.18	0.24	0.15	0.30	0.20	0.11	0.21	0.15
5-10	0.04	0.18	0.23	0.12	0.27	0.20	0.14	0.21	0.15
10-15	0.04	0.18	0.25	0.07	0.29	0.24	0.12	0.20	0.14
15-20	0.03	0.24	0.23	0.07	0.29	0.22	0.13	0.20	0.14
20-25		0.11	0.22	0.06	0.28	0.22	0.11	0.19	0.13
25-30			0.22	0.06	0.30		0.11	0.19	0.13
30-35			0.22	0.06	0.28		0.11	0.19	0.13
35-40			0.21	0.05	0.25		0.10	0.20	
40-45			0.22	0.06	0.24		0.10	0.19	
45-50			0.23	0.06	0.27		0.10	0.19	
50-55			0.24	0.06	0.25				
55-60			0.23		0.26				
60-65			0.24		0.25				
65-70			0.24						
70-75			0.24						
75-80			0.26						

Table 26. The MgO content (in %) of lake sediments from the Madeira River basin, Rondônia State, Brazil. According to Vergotti (2002)

Depth (cm)	Samuel	Paca	Araçá	Brasileira	Tucunaré	Nazaré	Conceição	Santa Catarina	Demarcação
0-5	0.03	0.57	1.45	0.68	1.29	1.39	1.20	1.45	1.03
5-10	0.04	0.58	1.43	0.72	1.26	1.38	1.39	1.48	1.17
10-15	0.04	0.51	1.42	0.32	1.36	1.42	1.26	1.44	1.20
15-20	0.03	1.25	1.47	0.38	1.36	1.43	1.31	1.44	1.18
20-25		0.27	1.43	0.28	1.41	1.43	1.25	1.41	1.19
25-30			1.39	0.32	1.38		1.15	1.44	1.17
30-35			1.42	0.29	1.33		1.04	1.41	1.19
35-40			1.43	0.24	1.33		0.86	1.46	
40-45			1.40	0.22	1.34		0.87	1.44	
45-50			1.40	0.21	1.37		0.93	1.44	
50-55			1.45	0.22	1.26				
55-60			1.41		1.31				
60-65			1.41		1.25				
65-70			1.40						
70-75			1.23						
75-80			1.25						

Table 27. The MnO content (in %) of lake sediments from the Madeira River basin, Rondônia State, Brazil. According to Vergotti (2002)

Depth (cm)	Samuel	Paca	Araçá	Brasileira	Tucunaré	Nazaré	Conceição	Santa Catarina	Demarcação
0-5	0.02	0.03	0.04	0.04	0.05	0.04	0.02	0.03	0.06
5-10	0.02	0.03	0.04	0.03	0.04	0.04	0.03	0.03	0.08
10-15	0.02	0.02	0.04	0.02	0.04	0.04	0.02	0.03	0.10
15-20	0.02	0.03	0.04	0.02	0.05	0.04	0.02	0.03	0.10
20-25		0.02	0.04	0.02	0.05	0.04	0.02	0.03	0.10
25-30			0.04	0.02	0.05		0.02	0.03	0.09
30-35			0.03	0.02	0.05		0.02	0.03	0.09
35-40			0.04	0.02	0.04		0.02	0.03	
40-45			0.04	0.02	0.04		0.02	0.03	
45-50			0.04	0.02	0.05		0.02	0.03	
50-55			0.05	0.02	0.04				
55-60			0.04		0.04				
60-65			0.05		0.04				
65-70			0.04						
70-75			0.03						
75-80			0.03						

Table 28. The Fe_2O_3 content (in %) of lake sediments from the Madeira River basin, Rondônia State, Brazil. According to Vergotti (2002)

Depth (cm)	Samuel	Paca	Araçá	Brasileira	Tucunaré	Nazaré	Conceição	Santa Catarina	Demarcação
0-5	8.12	4.90	7.78	6.77	7.05	5.85	4.61	5.87	6.56
5-10	8.22	4.78	7.75	5.16	6.79	5.77	5.43	5.66	6.74
10-15	8.61	4.10	7.29	3.98	6.78	5.89	4.65	5.24	7.40
15-20	7.62	5.04	6.74	3.91	6.48	5.96	4.50	5.08	7.34
20-25		3.05	6.33	3.77	6.50	6.17	4.06	5.02	7.51
25-30			6.56	3.73	6.77		3.78	5.01	7.83
30-35			6.85	3.54	6.15		3.50	5.12	7.67
35-40			6.89	3.41	6.11		3.20	5.07	
40-45			6.71	3.22	6.32		3.31	4.97	
45-50			6.57	2.85	6.51		3.61	5.05	
50-55			6.61	2.97	6.21				
55-60			6.84		6.09				
60-65			6.61		5.96				
65-70			6,52						
70-75			4,98						
75-80			5,00						

Table 29. The TiO_2 content (in %) of lake sediments from the Madeira River basin, Rondônia State, Brazil. According to Vergotti (2002)

Depth (cm)	Samuel	Paca	Araçá	Brasileira	Tucunaré	Nazaré	Conceição	Santa Catarina	Demarcação
0-5	2.29	1.22	0.85	1.21	0.88	0.95	0.91	0.93	1.18
5-10	2.25	1.22	0.86	1.17	0.88	0.91	0.90	0.94	1.26
10-15	2.30	1.26	0.86	1.30	0.90	0.95	0.92	0.97	1.25
15-20	1.41	0.91	0.87	1.35	0.90	0.92	0.90	1.02	1.22
20-25		1.45	0.88	1.47	0.87	0.94	0.98	1.01	1.21
25-30			0.91	1.39	0.87		1.07	1.00	1.22
30-35			0.88	1.39	0.90		1.29	1.00	1.22
35-40			0.89	1.40	0.88		1.38	1.01	
40-45			0.89	1.40	0.92		1.40	1.03	
45-50			0.90	1.47	0.87		1.37	1.06	
50-55			0.88	1.44	0.89				
55-60			0.88		0.88				
60-65			0.92		0.90				
65-70			0.91						
70-75			0.90						
75-80			0.93						

Table 30. The P_2O_5 content (in %) of lake sediments from the Madeira River basin, Rondônia State, Brazil. According to Vergotti (2002)

Depth (cm)	Samuel	Paca	Araçá	Brasileira	Tucunaré	Nazaré	Conceição	Santa Catarina	Demarcação
0-5	0.21	0.18	0.23	0.20	0.19	0.15	0.13	0.15	0.24
5-10	0.18	0.18	0.21	0.16	0.18	0.13	0.14	0.13	0.16
10-15	0.17	0.16	0.19	0.16	0.18	0.14	0.12	0.11	0.21
15-20	0.16	0.12	0.16	0.15	0.16	0.14	0.10	0.11	0.19
20-25		0.15	0.16	0.15	0.17	0.15	0.09	0.10	0.20
25-30			0.18	0.15	0.17		0.08	0.10	0.19
30-35			0.18	0.15	0.15		0.08	0.10	0.18
35-40			0.17	0.15	0.16		0.08	0.10	
40-45			0.16	0.14	0.15		0.08	0.10	
45-50			0.16	0.14	0.16		0.09	0.10	
50-55			0.15	0.14	0.16				
55-60			0.17		0.15				
60-65			0.16		0.16				
65-70			0.17						
70-75			0.11						
75-80			0.12						

Table 31. The average composition of lake sediments from the Madeira River basin, Rondônia State, Brazil

Parameter	Unit	Brasileira	Paca	Araçá	Tucunaré	Nazaré	Santa Catarina	Conceição	Demarcação	Samuel
Hg	ng/g	115.36	131.23	42.98	40.74	32.75	74.84	59.43	107.16	157.17
SiO_2	%	52.81	53.75	58.70	62.38	60.35	59.59	56.58	53.20	44.29
TiO_2	%	1.36	1.21	0.89	0.89	0.93	1.00	1.11	1.22	2.06
Al_2O_3	%	27.72	23.87	20.37	18.48	20.09	21.00	22.32	22.58	29.98
Fe_2O_3	%	3.94	4.37	6.63	6.44	5.93	5.21	4.07	7.29	8.14
MnO	%	0.02	0.03	0.04	0.04	0.04	0.03	0.02	0.09	0.02
MgO	%	0.35	0.64	1.40	1.33	1.41	1.44	1.13	1.16	0.03
CaO	%	0.07	0.18	0.23	0.27	0.22	0.20	0.11	0.14	0.04
Na_2O	%	0.22	0.37	0.73	0.82	0.74	0.70	0.45	0.23	0.08
K_2O	%	1.99	2.23	3.23	2.99	3.24	3.36	2.81	2.13	0.03
P_2O_5	%	0.15	0.16	0.17	0.16	0.14	0.11	0.10	0.20	0.19
OM[1]	%	4.98	9.37	5.70	3.60	3.48	6.10	7.74	7.58	18.07
LOI[2]	%	11.40	13.20	7.65	6.28	6.87	7.40	11.38	11.76	15.14
Moisture	%	39.39	-	44.14	44.52	39.42	45.78	55.59	44.93	28.48
Bases[3]	%	1.87	2.49	4.89	4.86	4.95	4.96	3.53	3.07	0.17
R_2O_3[4]	%	16.82	14.13	14.48	13.63	14.28	14.40	13.58	14.99	18.96

[1]OM = organic matter; [2]LOI = loss on ignition; [3]Bases = $MgO+CaO+Na_2O+K_2O$; [4]$R_2O_3 = Al_2O_3+Fe_2O_3+TiO_2$

The data are from Vergotti (2002). Bases and R_2O_3 were calculated as described by Colman (1982)

The data in table 17 show that the Hg concentration in sediment lakes from Madeira River basin is much lower than the range of 65-1675 μg/g reported by Pfeiffer et al. (1989) for sediments collected at Teotônio waterfalls during the peak gold mining activities, as they involved the use of a large number of mechanical dredges operating simultaneously. However, the Hg content values for are about 8-40 times higher than the average value corresponding to 4.4 ppb for rocks occurring in the area (regional background). The lakes Samuel, Brasileira and Paca are directly influenced by Jamari River, and they exhibited average Hg concentration corresponding to 157, 115 and 131 ng/g, respectively. Despite the lake Araçá is situated at the Jamari River margins, it appears it is not directly influenced by their waters as the average Hg concentration in the sediments (43 ng/g) is three times lower than the others lakes. The lake Demarcação possesses an average Hg concentration in the sediments above 100 ng/g, which is relatively similar to that of the lakes close to Jamari River despite its more distant location from the principal areas of occurrence of *garimpos*. The lakes situated near the Madeira River margins exhibited lower Hg contents, i.e. 33 ng/g (lake Nazaré) and 75 ng/g (lake Santa Catarina). Therefore, the average Hg concentration in lakes close to Madeira River corresponds to 50 ng/g, whereas for its two tributaries is 128 ng/g that is a value about 2.5 times higher.

Intense weathering processes affecting the crystalline rocks cropping out in the studied area have been partially responsible for the generation of fine sediments transported by water and deposited in the Madeira River, and, consequently, it would be expected that similar Hg concentrations in the rocks and sediments occur there. Because this was not the case, it is plausible to suggest that the influence of Hg anthropogenic emissions that occurred in the past, during the gold mining activities on the Madeira River, still remain at the present days when such activities have substantially reduced and are practically absent now.

Significant correlation (r = 0.78) was found between the average Hg content and organic matter in the lake sediments, as shown in figure 18. Since the loss on ignition (LOI) (table 19) expresses the organic matter + adsorbed water + water in crystal lattices and fluid inclusions + CO_2 of carbonates + SO_2 of sulfides, it is a parameter related to organic matter, thus, justifying the significant relationship (r = 0.90) between the average Hg content and LOI in the lake sediments (figure 18).

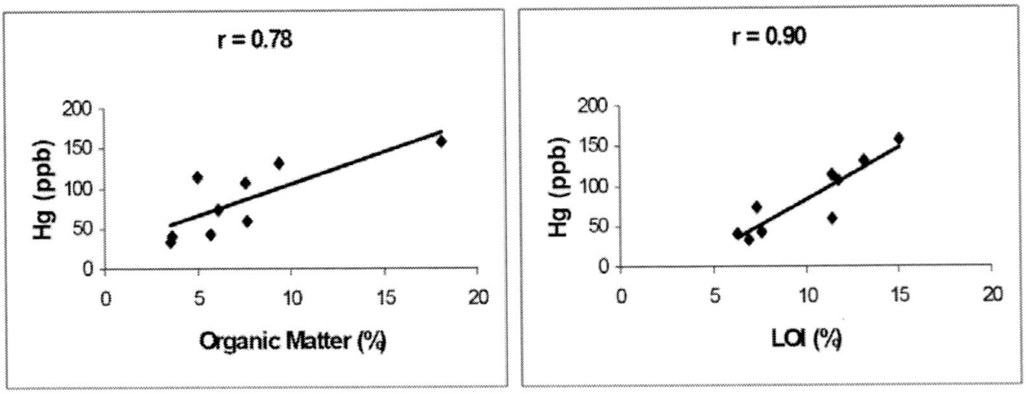

Figure 18. The relationships among Hg, organic matter, and loss on ignition (LOI) contents for lake sediments from Madeira River basin, Rondônia State, Brazil. According to Bonotto et al. (2006).

Beyond modifying the bulk density and porosity of soils and sediments, organic matter also influences their specific surface (area per mass, often expressed in m^2g^{-1}), since 1% of organic matter in porous matrices may cause an increase of the specific surface by about 7 m^2 (Kiehl, 1977). Consequently, other parameters are also affected significantly, for instance, the cation exchange capacity, the cation adsorption, and the percentage of water retention in soils/sediments, which, in general, increase with increasing specific surface area.

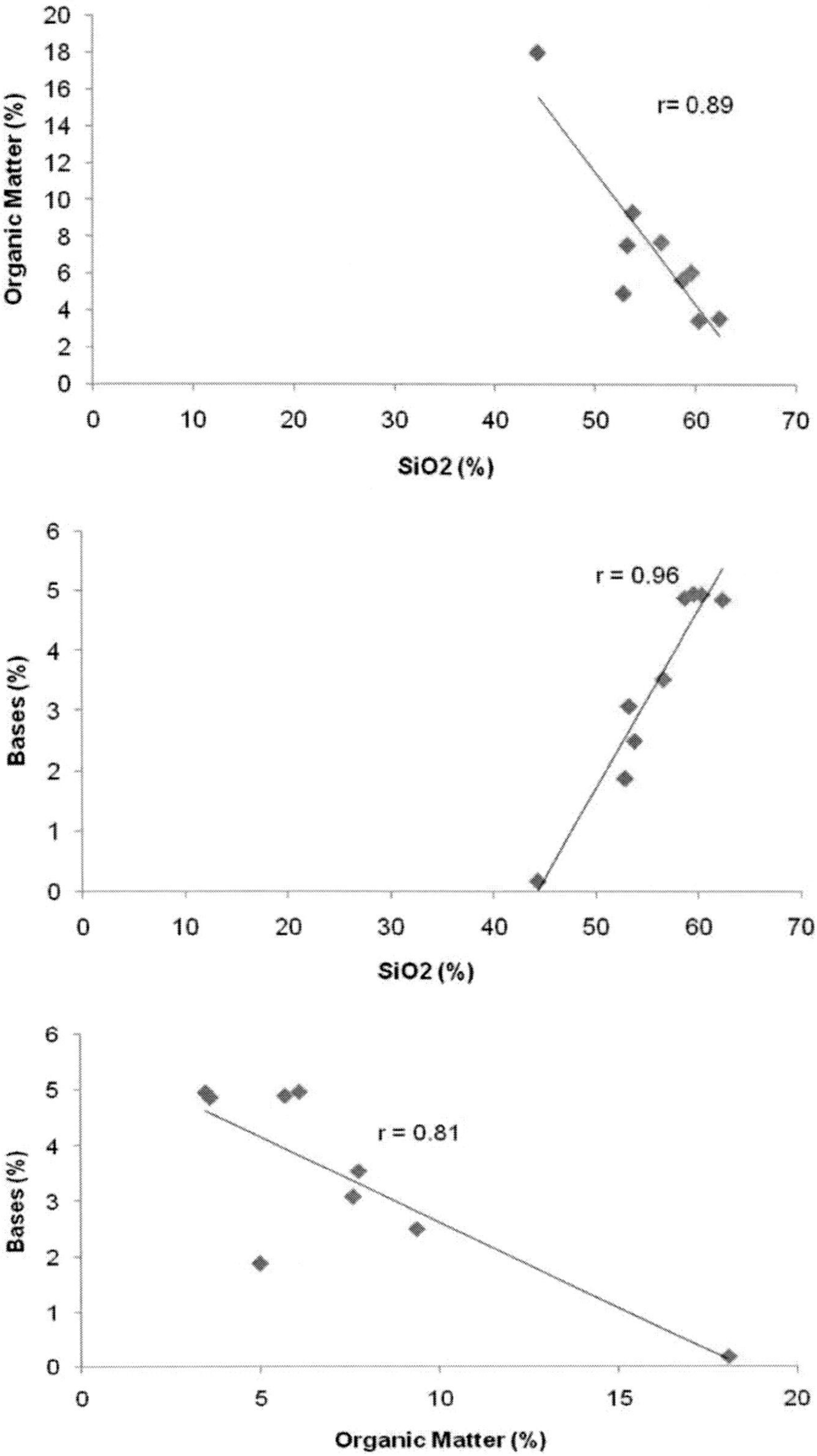

Figure 19. The relationships among SiO_2, Organic Matter and bases in the lake sediments from Madeira River basin, Rondônia State, Brazil. According to Vergotti (2002).

Silica is the major constituent in all lake sediments profiles (table 21) and is inversely related to organic matter (figure 19). Furthermore, because intense weathering processes occur in the area, it occurs the generation of the typical products of granitoids, i.e. feldspars forming kaolinite. Consequently, the bases are associated to silica, as justified by the significant correlations between silica and bases, and bases and organic matter (figure 19).

The silica decreasing causes an increase on the organic matter content and raising of the specific surface of the sediments, consequently, increasing the exchange/adsorption capacity of the major oxides resistant to weathering, which, like mercury, are significantly correlated with organic matter (figure 20). Therefore, the Hg in the lake sediments is also associated to heavy metals Al, Fe and Ti due to the organic matter influence, as indicated the significant correlation between R_2O_3 and Hg (figure 20). The non-association of Hg to silica and bases is evidenced by the inverse correlations among these parameters (figure 20).

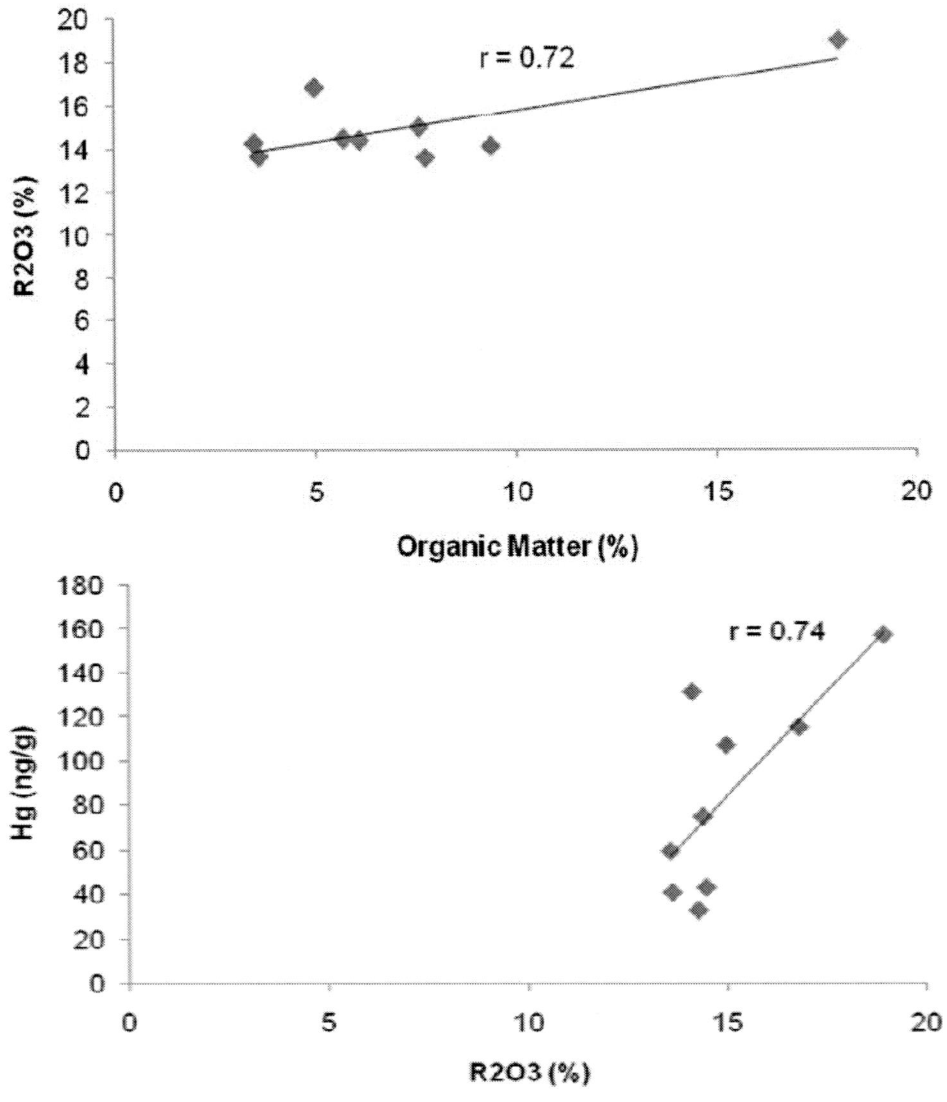

Figure 20. Continued on next page.

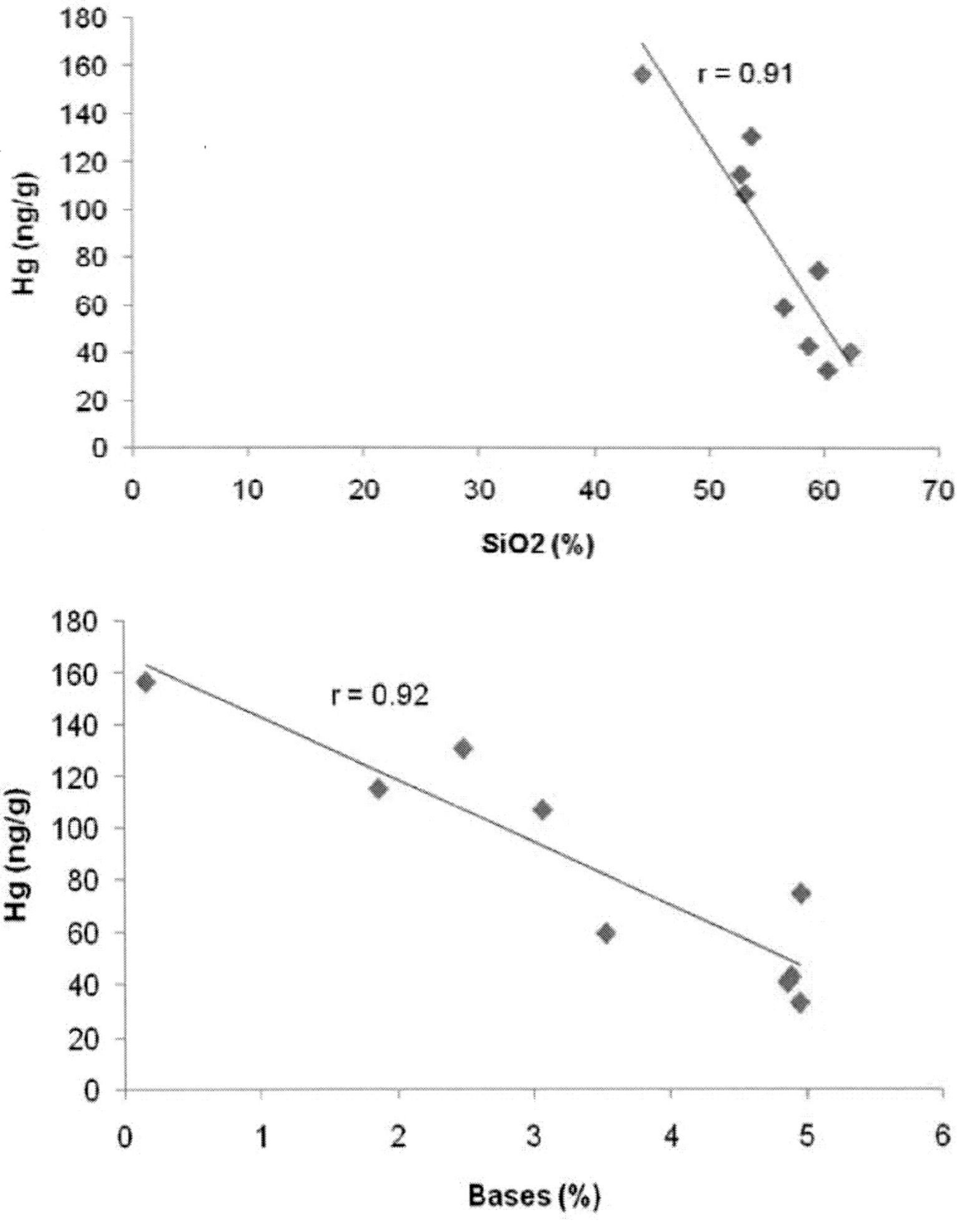

Figure 20. The relationships among organic matter, R_2O_3, bases and Hg in the lake sediments from Madeira River basin, Rondônia State, Brazil. According to Vergotti (2002).

In terms of the variability of the Hg content relatively to the depth of the profiles, there are significant differences if the average values are considered (table 31). It was only identified significant correlation between the Hg content and depth for the sediments profile from lake Brasileira (figure 21). In this core, it was also verified significant correlations between Hg and LOI, Hg and R_2O_3, Hg and SiO_2 and Hg and bases (figure 21).

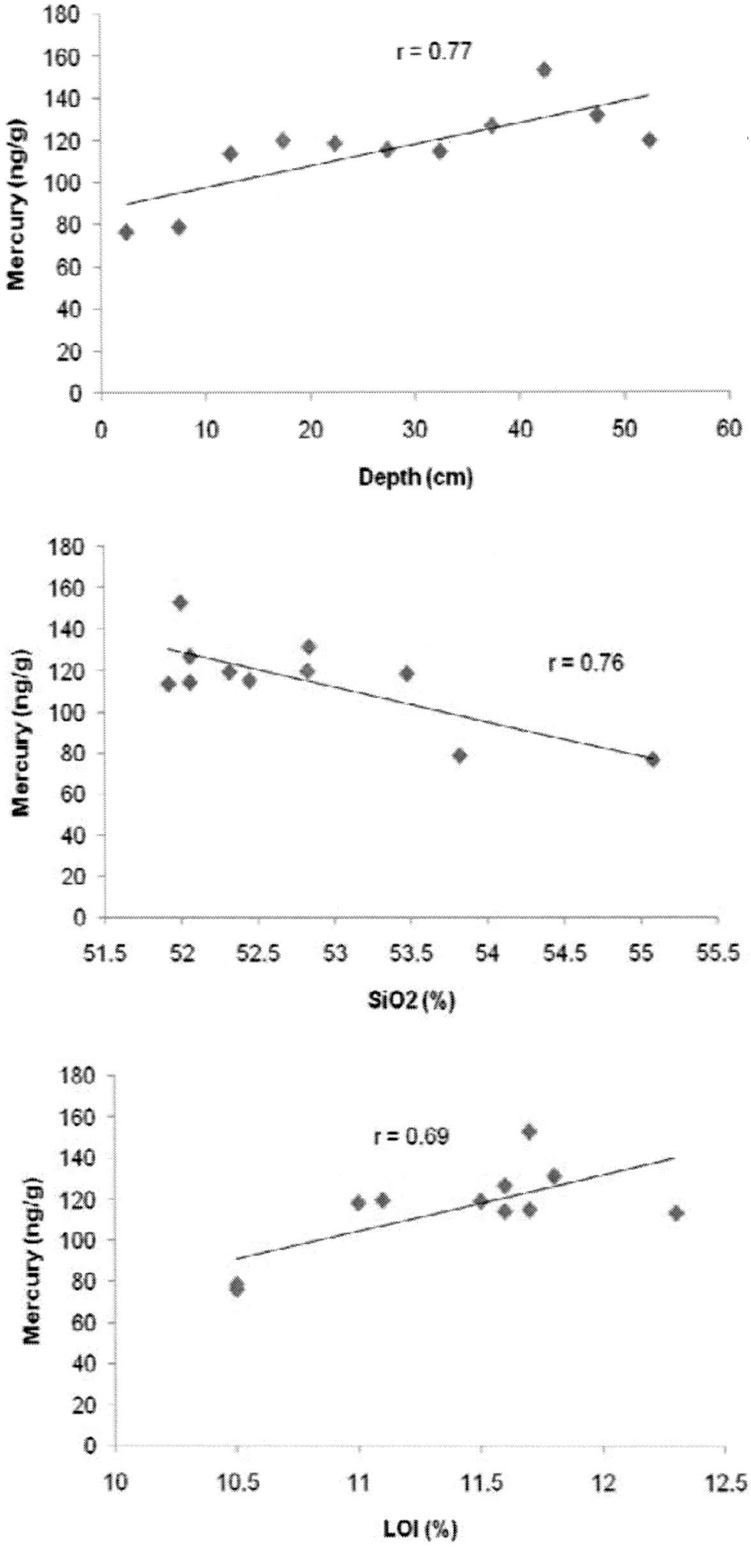

Figure 21. Continued on next page.

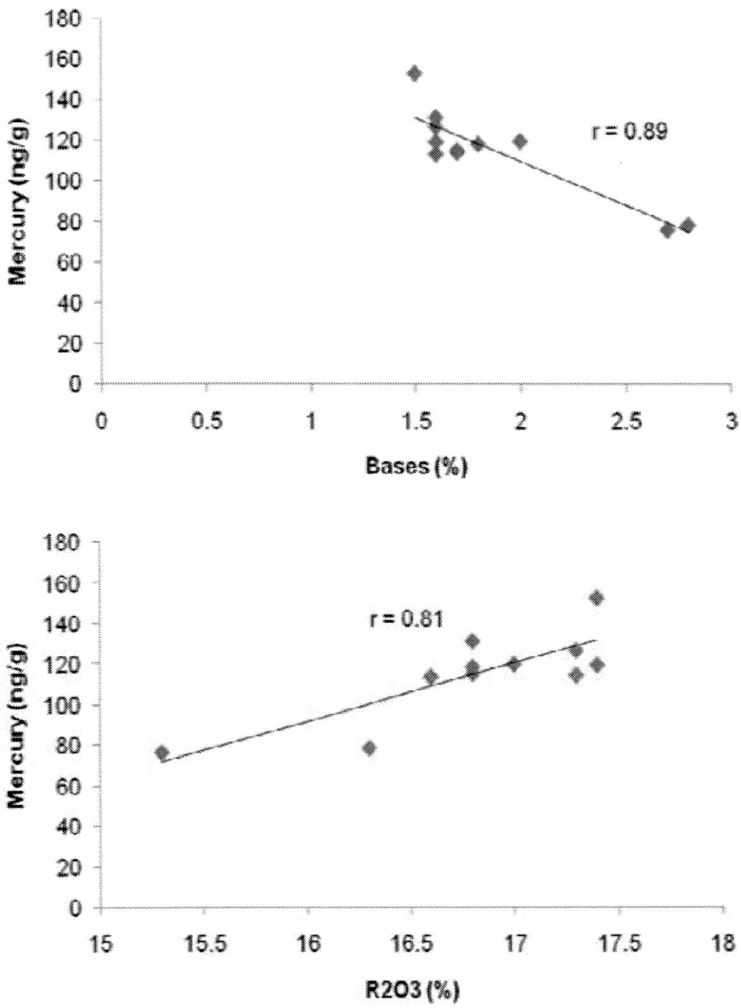

Figure 21. The relationships among depth, LOI, R_2O_3, SiO_2, bases and Hg in the sediments from lake Brasileira at the Madeira River basin, Rondônia State, Brazil. According to Vergotti (2002).

Chapter 10

The Association of Mercury and Selenium in Sediments from Madeira River Basin

The development of farmlands in Brazilian Amazon is a recent issue that has caused several debates and discussions. It has been important to provide information about the presence of heavy metals in fertilizers utilized in agriculture as its content in soils and grasses of cattle farms have induced variable degree of intoxication on animal and human population. Thus, it is another possible source for mercury inputs in aquatic systems in addition to the gold mining activities, and because seleniferous agricultural drainage wastewater has become a new major source of pollution in the world, this chapter describes the results obtained in the chemical analyses of selenium in sediments collected at the Madeira River basin, Rondônia State, with the purpose of investigating some possible associations with the mercury retention.

Major Facts About the Se Presence in the Environment

Selenium was discovered in 1817 by Berzelius and Gahn who used copper pyrites as a source of SO_2 for the H_2SO_4 manufacture and generated a residue containing a new element that was named from the Greek σελήνη, the moon (Bagnall, 1966). In present days, it is used in photocells, xerography, metal rectifiers, for the dehydrogenation of hydroaromatic compounds valuable in elucidating the structure of several natural products, as an oxidizing agent and catalyst in organic syntheses, as oxidation inhibitors in lubricating oils, in shampoos as a dandruff inhibitor, as bases for phosphors, for red ceramic and enamel pigments, for coloring glass red or reddish yellow and to neutralize the greenish tints caused by the presence of iron in certain types of glass in railway, marine and other signal lamps.

Selenium is necessary as a trace nutrient to man, but it may also cause toxic effects when ingested in non-appropriate amounts. For instance, hydrogen selenide when breathed in air, even at very great dilutions, causes headache and nausea, and, at high concentrations, acute irritation of the mucous membrane; many selenium compounds cause eczema and inflammation on contact with the skin, being absorbed essentially in the kidneys, spleen and liver when ingested, and only slowly released from the body, causing the subject to have extremely offensive smelling breath and perspiration for a long period (Bagnall, 1966). The selenium deficiency and toxicity, mainly in the field of animal nutrition, has been considered elsewhere (Rosenfeld and Beath, 1964; Kabata-Pendias and Pendias, 1984), with the problem of seleniferous soils and their effects on animal and human population being reported in Colombia since colonial times (García et al., 1999). Seleniferous agricultural drainage wastewater has become a new major source of pollution in the world, where, in the USA, large areas of farmland in 17 western states generate contaminated drainage with Se concentrations much higher than the US Environmental Protection Agency water-quality criterion for the protection of aquatic life (Kharaka et al., 1996).

Thus, environmental problems involving generated natural and anthropogenic selenium are becoming widespread, being necessary to enhance the knowledge of its behavior in different situations. Selenium predominantly occurs in four forms in the natural environment. The highly soluble selenate (SeO_4^{2-}) is found in alkaline oxidizing environments. In more acidic environments, selenium tends to occur as selenite (SeO_3^{2-}) ion, which is not as soluble as selenate, whereas in more reducing environments, selenium may occur as the metal (Se^0) or the selenide (Se^{2-}) ion (McNeal et al., 1990). The types of reactions that affect selenium in sediments can be both biotically and abiotically controlled, involving conversions between particulate and dissolved phases and including redox reactions able to change the oxidation state of selenium (Velinsky and Cutter, 1991).

Because the development of farmlands in Brazilian Amazon is a recent issue that has caused several debates and discussions, it is important to provide information about the selenium behavior at the Madeira River basin, since its content in soils and grasses of cattle farms have induced variable degree of intoxication on animal and human population (García et al., 1999). Selenite ion, Se(IV), was chosen to investigate the selenium occurrence in the area, due to the dominance of acid conditions there.

SAMPLING AND ANALYTICAL PROTOCOL

The Se(IV) distribution data was performed for two types of materials: 1) sediments collected at 12 points along 25 km in the Madeira River between the Teotônio and Santo Antônio waterfalls, upstream from Porto Velho (area of about 260 km^2) (figure 8); this sector of the river is not suitable for navigation, and fieldwork was performed from a small boat during the dry season in a two-week period in August 1996. Approximately 500 g of sediments were collected from each sampling point with a bottom dredge that slowly trawled upstream and transferred samples to a half-filled thick polyethylene bag. 2) core sediments (20 to 80 cm deep) collected at nine lakes between Jamari and Ji-Paraná rivers, that are tributaries of Madeira River; the studied area is situated between 8° and 9°S and between 62° and 64°W (figure 17), being the fieldwork realized during October 1999 and February 2000, also utilizing a small boat that crossed several *igarapés* in order to have access to the lakes. A 1-m long acrylic tube (10 cm in diameter) adapted to an iron outliner was driven into the sediments preferentially at the middle of each lake, being each profile cut in 5-cm thick slices that also were transferred to polyethylene bags.

All bags containing the sediments were stored in iceboxes and transported to the laboratory. The material was dried at 60°C, crushed, quartered, sieved for separation of the < 200 mesh (< 74 μm) size fraction, and divided into aliquots for different analyses. The determination of Se(IV) was performed in approximately 0.5 g of dried material, that was immersed in 1 mL of high purity distilled water, being added 2 mL of a 3:1 HCl-HNO3 mixture for utilizing the acid digestion procedure through heating at 50°C during 15 minutes (Cutter, 1985). The solution was separated from the sediments by filtration, being its volume elevated up to 100 mL with pure distilled water. Standard analytical techniques (Hach, 1992) were used for treating the solution in order to obtain its selenium concentration by spectrophotometry. Diaminobenzidine reacted with all selenium present as selenite (Se^{4+}) to give a yellow-colored piazselenol complex extracted with toluene, being its color read by Hach DR/2000 spectrophotometer at 420 nm (Hach, 1992). Table 32 reports all obtained Se(IV) data.

Table 32. Se(IV) content (in µg/g) of sediments from a) Madeira River between Teotônio and Santo Antônio falls and b) lakes. The data are reported by Costa Neto et al. (2003)

	Sampling Point No.	1	2	4	5	7	9	10	11	13	15	16	17	Mean
(a)	Value	12	6	2	2	<1	2	2	6	4	6	6	8	4.8

	Depth Range (cm)	Samuel	Paca	Araçá	Brasi-leira	Tucu-naré	Nazaré	Concei-ção	Santa Cata-rina	Demar-cação
(b)	0-5	6	16	2	2	16	4	12	4	18
	5-10	2	16	2	6	22	6	10	2	4
	10-15	8	16	2	6	18	2	18	6	2
	15-20	4	4	<1	12	8	6	10	6	2
	20-25		10	2	8	8	8	4	14	4
	25-30			4	10	14		2	6	2
	30-35			4	14	10		4	4	10
	35-40			8	16	4		2	6	
	40-45			4	14	6		6	8	
	45-50			4	8	6		2	6	
	50-55			2	18	2				
	55-60			6		2				
	60-65			4		2				
	65-70			4						
	70-75			6						
	75-80			10						
	Mean	5.0	12.4	4.1	10.4	9.1	5.2	7.0	6.2	6.0

The Importance of Weathering Processes

Rock weathering processes often cause chemical changes in soils and sediments. In lateritic soils, silica may be eliminated from silicates, forming free Al-Fe hydroxides (Erhart, 1973). Intense weathering conditions favor the occurrence of laterisation, enhancing the Al-Fe content and diminishing the Si concentration in relation to simply kaolinized rock matrices (Schellmann, 1980). In general, the evolution of laterisation processes is related to the nature of the original rock matrix and involves all Al-silicates rock types, with their occurrence affecting very intensively the crystalline rocks of the studied area. The chemical composition of the analyzed lake sediments reflects these aspects, as demonstrates the generalized inverse significant correlation between SiO_2 and Al_2O_3 (figure 22). Thus, Si-poorer sediments are more Al-enhanced, suggesting the occurrence of more prolonged/intense weathering processes, with the opposite also being probable.

Al_2O_3 is a constituent often used to identify gains and losses during chemical weathering (Colman, 1982; Faure, 1991). Such oxide is assumed to remain constant in soil profiles, despite its variability in concentration against depth (Faure, 1991). SiO_2/Al_2O_3 ratios have been utilized by geochemists as weathering indices (Pedro, 1968; Pedro and Sieffermann, 1979), with the highest values indicating less prolonged/intense processes. Each sediments profile plotted in figure 22 exhibited different SiO_2/Al_2O_3 ratio, thus, characterizing more/less prolonged/intense weathering processes. For instance, SiO_2/Al_2O_3 ratios corresponding to 3-4.1 and 4.6-5 were found, respectively, for Brasileira and Santa Catarina lake sediments. The mean SiO_2/Al_2O_3 ratio of each sediments profile was: Samuel=2.5, Paca=3.8, Araçá=4.9, Brasileira=3.2, Tucunaré=5.7, Nazaré=5.1, Conceição=4.3, Santa Catarina=4.8, and Demarcação=4.

The Presence of Se(IV) at the Madeira River Basin

The mean Se(IV) content (4.8 μg/g) of sediments between Teotônio and Santo Antônio falls is about 100 times higher than the average Hg content (0.047 μg/g) (table 10), with non-significant correlation being found between Se(IV) and other parameters analyzed for these sediments (table 11). The average Se(IV) content of most of the core lakes (4-7 μg/g) is practically equivalent to the value found for sediments from Madeira River, whereas those determined for Paca, Brasileira, and Tucunaré lakes (9.1-12.4 μg/g) correspond to the highest values reported for soils close to the border of streams recognized as carriers of seleniferous sediments in Colombia (García et al., 1999).

Partial hydrolysis developed under slow flow conditions may occur in several intensity degrees during weathering in tropical regions characterized by hot and wet climate. X-ray diffraction of sediments from Madeira River between Teotônio and Santo Antônio falls identified quartz, mica, feldspars, pyrophyllite, gibbsite, ilmenite, amorphous Fe-Mn oxides, kaolinite and chlorite-vermiculite mixed layers, and some of these minerals represent alteration products of granitoids (i.e. feldspars yield kaolinite). The presence of kaolinite was also confirmed on the study of the stability relations of mineral phases in the systems $K_2O(CaO, Na_2O)$-Al_2O_3-SiO_2-H_2O (figure 11).

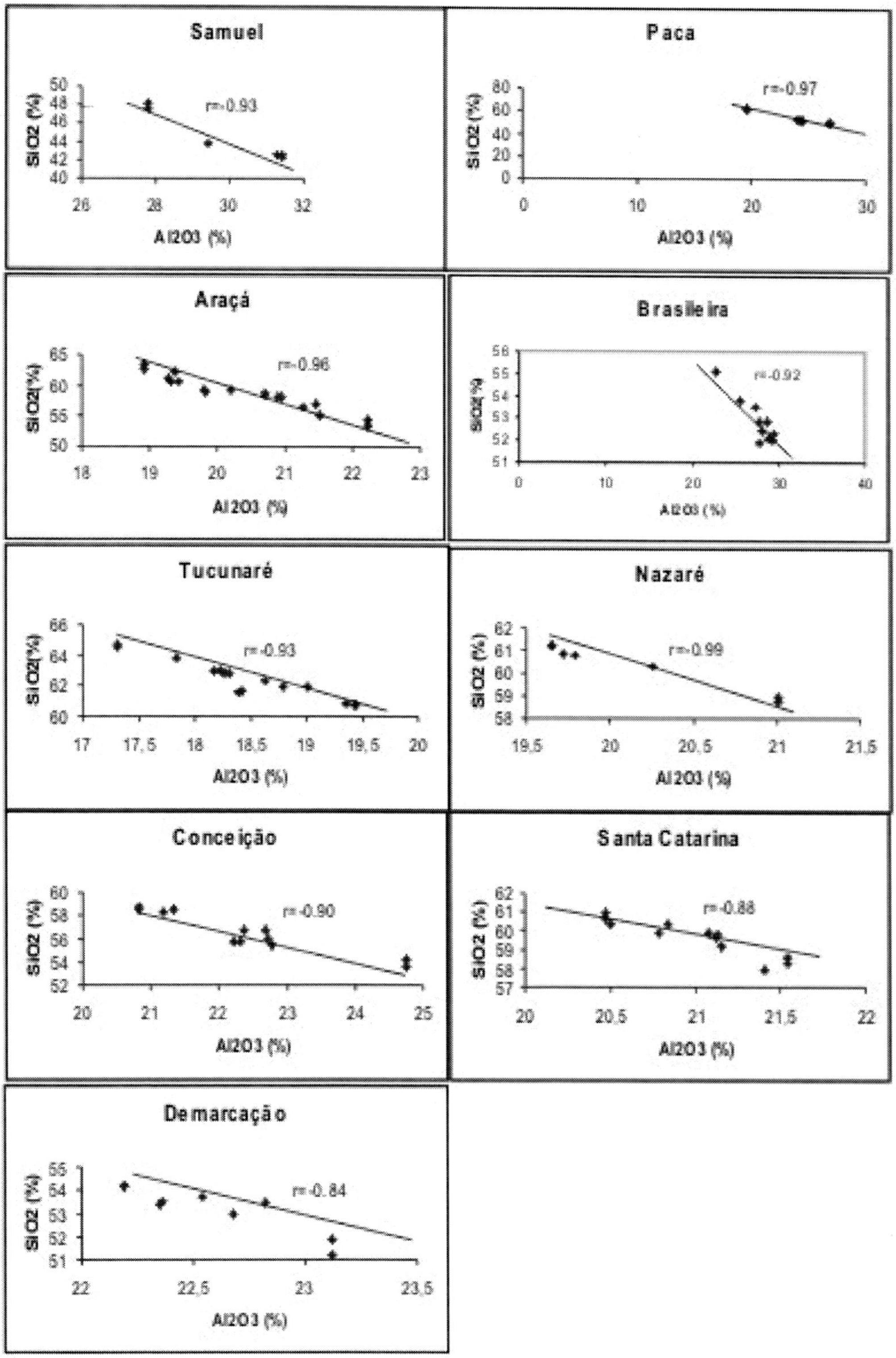

Figure 22. Relationship between SiO_2 and Al_2O_3 for core sediments of lakes at the Madeira River basin, Rondônia State, Brazil. The data are given in tables 21 and 22 as reported by Vergotti (2002).

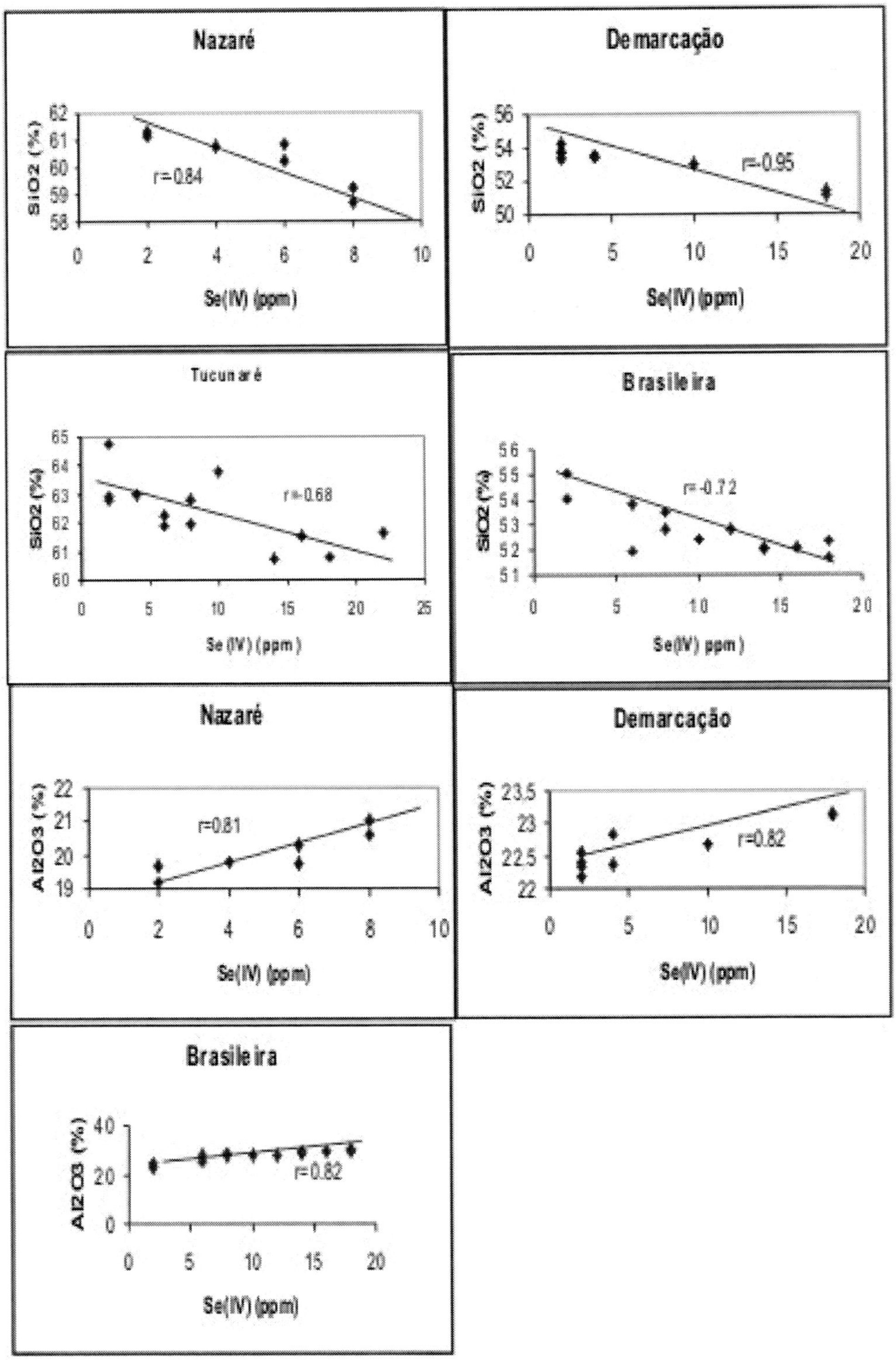

Figure 23. SiO_2-Al_2O_3 vs. Se(IV) relationships for core sediments of lakes at the Madeira River basin, Rondônia State, Brazil. The data are given in tables 21 and 22 as reported by Vergotti (2002).

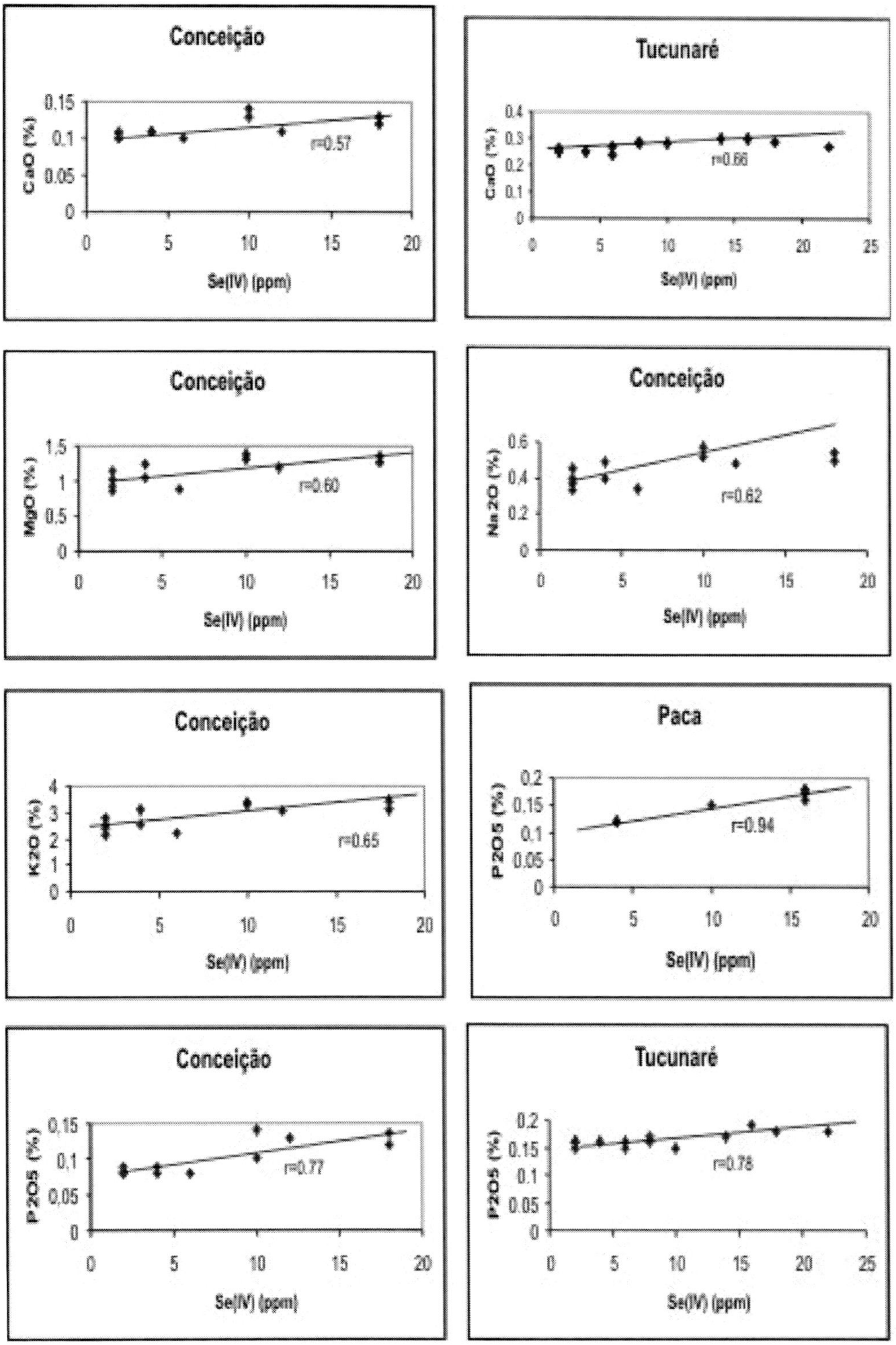

Figure 24. CaO-MgO-Na_2O-K_2O-P_2O_5 vs. Se(IV) relationships for core sediments of lakes at the Madeira River basin, Rondônia State, Brazil. The data are given in tables 23-26 and 30 as reported by Vergotti (2002).

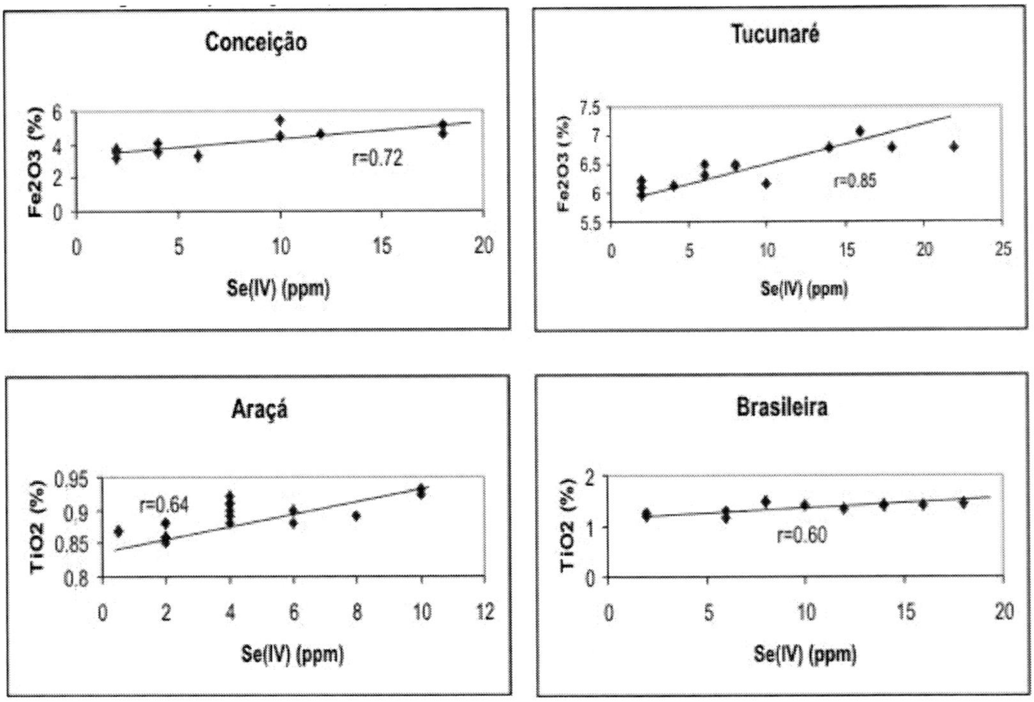

Figure 25. Fe_2O_3-TiO_2 vs. Se(IV) relationships for core sediments of lakes at the Madeira River basin, Rondônia State, Brazil. The data are given in tables 28 and 29 as reported by Vergotti (2002).

An aspect frequently emphasized in literature (Velinsky and Cutter, 1991; Kharaka et al., 1996) is that the Se(IV) selenite ion, as $HSeO_3^-$ and SeO_3^{2-}, exhibits a great tendency to be absorbed by clay minerals and Fe oxi-hydroxides. Several direct significant relationships reflecting this aspect were found when the chemical data reported in tables 21-29 are correlated with the measured Se(IV) content, as plotted in figures 23-25, i.e. Al_2O_3-CaO-MgO-Na_2O-K_2O-TiO_2-Fe_2O_3 vs. Se(IV). The inverse significant correlation between SiO_2 and Se(IV) (figure 23) is related to the general inverse relationship between SiO_2 and Al_2O_3 (figure 22).

Beyond the typical Al_2O_3-Fe_2O_3 accumulations during chemical weathering, other constituents also are present in resistates/minor refractory minerals more resistant to weathering (such as apatite, zircon, and sphene, among others) that are incorporated intact in the coarser grained fraction of the analyzed sediments, becoming additional sorbents for selenite ion. A good example for this is the direct significant relationship between P_2O_5 and Se(IV) for core sediments from Paca, Conceição, and Tucunaré lakes, as plotted in figure 24 (the P_2O_5 data are given in table 30).

The Retention of Selenium and Mercury at the Madeira River Basin

Organic matter has been often considered an important complexing agent for selenite ion (Velinsky and Cutter, 1991; Kharaka et al., 1996), and a direct evidence for this was found for sediments of Demarcação lake (figure 26). Since the Loss on Ignition (LOI) expresses the organic matter + adsorbed water + water in crystal lattices and fluid inclusions + CO_2 of carbonates + SO_2 of sulfides, it is a parameter related with organic matter, thus, justifying the significant relationships between LOI and Se(IV) plotted in figure 26.

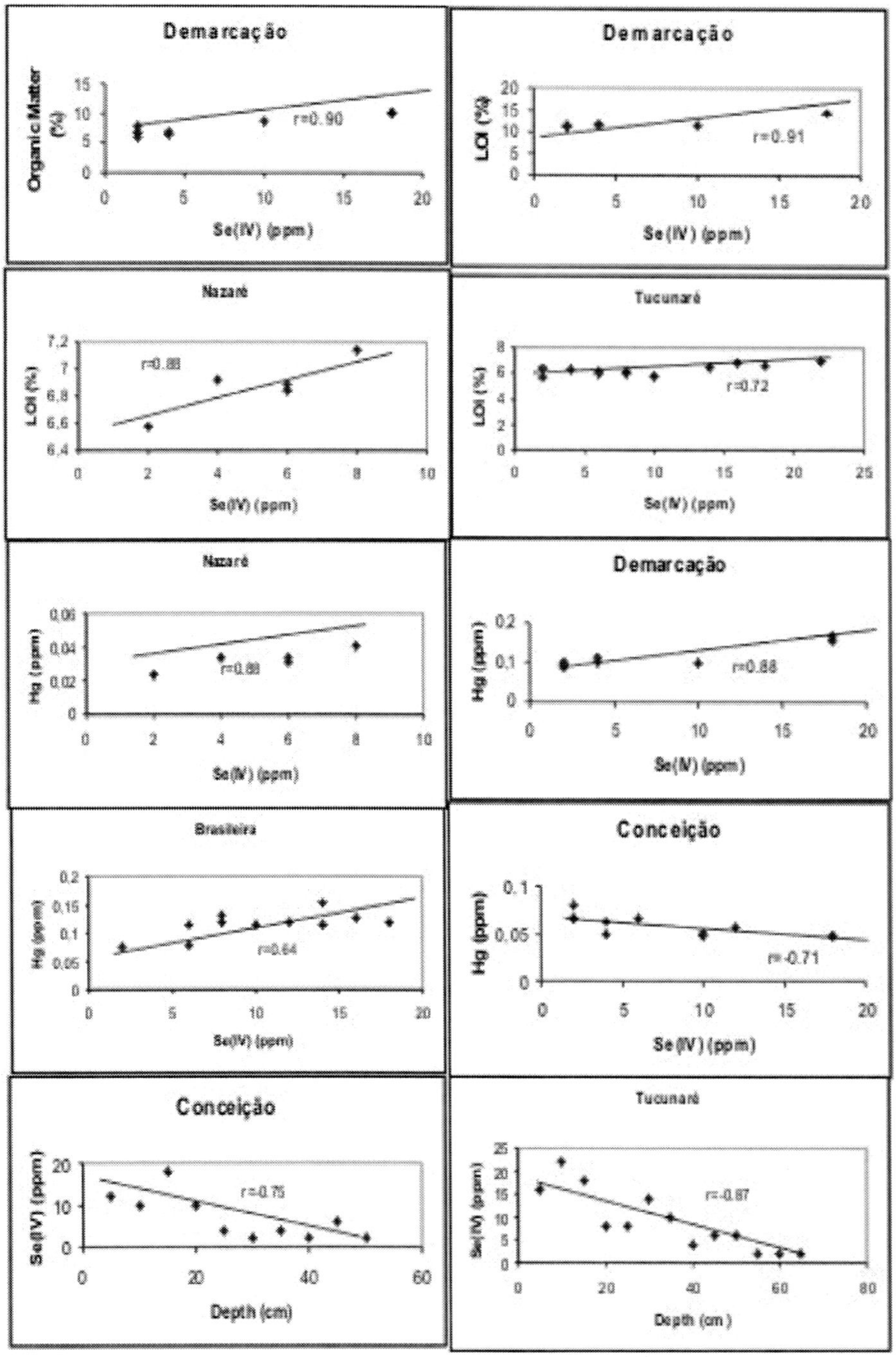

Figure 26. Organic Matter-LOI-Hg vs. Se(IV) relationships for core sediments of lakes at the Madeira River basin, Rondônia State, Brazil. The data are given in tables 17-19 as reported by Vergotti (2002).

The Hg absorption by organic matter was identified on the lake sediments (figure 18), so that the significant relationship between Hg and Se(IV) plotted in figure 26 for core sediments of Nazaré and Demarcação lakes is certainly related to the organic matter influence on the heavy metals absorption. However, in the case of core sediments of Brasileira lake, the significant relationship between Hg and Se(IV) (figure 26) may be justified by the presence of clays, amorphous Fe-Mn oxi-hydroxides, and minor refractory minerals, rather than organic matter. An opposite trend between Hg and Se(IV) was found for core sediments of Conceição lake, being related to the contrary behavior of these constituents with the core depth, i.e. Hg increases with depth (Vergotti, 2002), whereas Se(IV) decreases with depth (figure 26); neither Hg nor Se(IV) are correlated with organic matter in this sediments profile. Another significant relationship expressing the decrease of Se(IV) with depth was verified for core sediments of Tucunaré lake (figure 26).

Table 33. Significant correlation coefficient among Se(IV) content (μg/g) and other parameters evaluated for lake sediments of the Madeira River basin

	Lake					
Parameter	Conceição	Santa Catarina	Demarcação	Tucunaré	Araçá	Brasileira
Depth					0.68	0.78
SiO_2		0.64			0.62	
TiO_2	-0.68		-0.70			
Al_2O_3					-0.58	
Fe_2O_3					-0.64	-0.74
MnO			-0.83			-0.67
MgO		-0.61	-0.85		-0.66	-0.68
CaO		-0.57				-0.72
Na_2O				-0.73		-0.68
K_2O			-0.88			-0.64
P_2O_5					-0.59	-0.72
LOI		-0.73				

The Se mobility in the environment is dependent of biogeochemical, physical and chemical parameters that are expressed, for instance, by temperature, pH, redox potential, dissolved oxygen and sulfur in the water column, phytoplankton composition, bioturbation occurrence, and biological activity intensity, among others (Velinsky and Cutter, 1991). Despite their importance on the Se(IV) migration, they have not been properly evaluated at the Madeira River basin. Therefore, some of them possibly may be important/responsible for explaining the significant relationships reported in table 33 for several lakes of the Madeira River basin, which are opposite to those plotted in figures 22-26.

Chapter 11

MERCURY IN FISHES FROM MADEIRA RIVER BASIN

The gold exploitation in several areas of the Brazilian Amazon has caused environmental problems due to the presence of mercury in the biosphere, whether in liquid, ionic or vapour form. Special attention has been given to mercury inputs at the Madeira River in the Amazon area by numerous investigators, where the contamination of the aquatic food-chain has been demonstrated for several piscivorous fishes that exhibited high mercury concentrations. This chapter describes the results obtained in the chemical analyses of mercury in a large number of fishes collected at the Madeira River basin, Rondônia State, with the purpose of investigating their possible contamination.

SAMPLING AND ANALYTICAL PROTOCOL

The evaluation of the mercury presence in fishes of the Madeira River basin is important due to the possible ingestion of this metal in food. The Hg concentration has been determined for 419 samples of fishes from Madeira, Jaciparaná, and Jamari rivers in the Madeira River basin between 7 and 11° parallels south and between 62 and 65° meridians west in Rondônia State, Brazil (figure 1). The samples were obtained between April 1994 and January 1998 in the central market located at the city of Porto Velho, as well collected directly from the rivers after using roads and small boats, because the selected area is not suitable for navigation. The specimens were identified according to their popular name during the fieldwork. Afterwards, they were stored in polyethylene bags that were kept in iceboxes and transported up to the laboratory.

The weight and length of the specimens were recorded at the laboratory according to standard procedures (Santos, 1991). It was also indicated if their skin contains scales or not, as well their possible scientific name based on criteria applied to Amazonian fishes (Santos, 1991). Then, a fillet was cut from the back of the specimen and maintained under freeze (-5 °C) up to the time of analysis.

Digestion and extraction procedures were used for Hg analysis in the samples. About 0.5 g of each sample was weighed in Teflon beakers/tubes, 1 mL of 30% H_2O_2 was added plus 2 mL of a 1:1 H_2SO_4-HNO_3 mixture and 5 mL of 5% $KMnO_4$. This material was inserted in a

630 W microwave oven and maintained there (temperature~55°C) during about 35 minutes. Alternatively, digestion was also performed in an opened system. Afterwards, the solution was brought to room temperature (~25°C) and the excess $KMnO_4$ was neutralized with some drops of 12% hydroxylamine hydrochlorate. The volume was raised up to 10 mL with ultra pure water prepared on a Milli Q system that uses ion exchange resin and Millipore 0.22 μm filter.

The analyses of total Hg concentrations were done using atomic absorption spectrophotometers with cold vapor generation. The equipment utilized is a Perkin Elmer model FIMS-400 atomic absorption spectrophotometer, containing a cylindrical cell with quartz window, a Hg lamp (253.7 nm), a specific detector for absorption on the Hg wavelength, a system of flow injection analysis, and a cold vapor generator system. The parameters of calibration utilized in this equipment consisted on an Ar flow of 50 $mL.min^{-1}$, a 3% HCl flow of 9-11 $mL.min^{-1}$, and a 0.2-0.5% $NaBH_4$+NaOH flow of 5-7 $mL.min^{-1}$. The concentrations of the analytical solutions for obtaining the calibration curve of the equipment varied between 2 and 40 $\mu g.L^{-1}$, and were prepared from a 2 $\mu g.mL^{-1}$ stock solution preserved with 5% HNO_3 + 0.01% $K_2Cr_2O_7$. Each aliquot representing the sample (0.5 g) was divided in three sub-aliquots. All analytical determinations were made in triplicate for each sub-aliquot, yielding, therefore, nine measurements for each sample. The final result for each sample corresponded to the mean value of the nine readings.

The Hg Content in Selected Fishes

Table 34 reports the Hg content in the analyzed fishes. The average values of the Hg content in the analyzed fishes are greatly variable, ranging from 0.01 to 1.86 μg/g. The highest value (1.86 μg/g) was found for coroatá (*Platynematichthys notatus*) (figure 27) whose main diet consists on the ingestion of small fishes and organic matter and microorganisms associated to mudstone at the bottom of lakes and margins of rivers (Urenha Jr., 2008). The species sampled are widely spread over the area, being commonly consumed by the population. However, considering the number of specimens sampled per specie, the economic importance and the use in alimentation, four types were selected for more careful evaluation of the data obtained. They are: *Prochilodus* sp., *Serrasalmus* sp., *Hoplias malabaricus*, and *Cichla ocellaris*. A summary of the parameters obtained for these fishes is shown in table 35.

The WHO safety limit (WHO, 1976) for ingestion of mercury in fishes is 0.5 μg/g and considering the range of values reported in table 35, only the specimens of *Prochilodus* sp. do not reach the maximum allowable limit in any analyzed sample. These fishes ingest organic matter and microorganisms associated to mudstone at the bottom of lakes and margins of rivers, realizing long migration for reproduction. *Serrasalmus* sp., *Hoplias malabaricus*, and *Cichla ocellaris* are classified as carnivorous fishes, and among them the average Hg content is higher in *Serrasalmus* sp. (0.8 μg/g) that exceeds the WHO safety limit of 0.5 μg/g.

However, such contamination level is lower than that found in other piscivorous fishes in the Amazon area, since values exceeding by almost four times the WHO safety limit have been reported (Maurice-Bourgoin et al., 2000). The same relationship between the Hg content and the alimentation pattern of the specie had been identified in previous studies developed in

Brazil, with the higher Hg contents occurring in individuals of higher trophic levels (Pfeiffer et al., 1988; Lacerda et al., 1989; Malm et al., 1990; Malm, 1991).

STATISTICAL ANALYSIS OF THE HG CONCENTRATION DATA

The Hg content data obtained for *Prochilodus* sp., *Serrasalmus* sp., *Hoplias malabaricus*, and *Cichla ocellaris* were submitted to statistical analysis, as reported in table 36. All data adjust to a normal distribution (figure 28), inclusive those obtained for *Cichla ocellaris* that indicate a bi-modal normal distribution.

Statistical tests of correlation were performed between weight and length and between the Hg content and weight for the data obtained for *Prochilodus* sp., *Serrasalmus* sp., *Hoplias malabaricus*, and *Cichla ocellaris*. Figure 29 plots all relationships found among these parameters, where it is possible to see that the weight of the specimens is directly related to their length, as expected. The Hg content raises in accordance with the weight of the carnivorous fishes *Serrasalmus* sp., *Hoplias malabaricus*, and *Cichla ocellaris* (figure 29), evidencing the cumulative effect of mercury in tissues, i.e. the increase in weight is normally related to the age of the specimen, that causes a longer assimilation of Hg present in food, not in the metallic form, but bio-available in the ionized (organic) form. Beyond the Hg content values in the non-piscivorous fishes *Prochilodus* sp. not exceed the WHO safety limit of 0.5 μg/g, the Hg content and weight in these fishes are not related (figure 29), reinforcing the importance of the alimentation habit on the Hg bioaccumulation.

Table 34. The average Hg content in fishes from Madeira River basin, Brazil. According to Gali (2002)

Popular name	Scientific name	N*	Hg (μg/g)
Acará	*Geophagus* sp.	10	0.09
Acari Bodó	*Lipsarcus pardalis*	1	0.04
Apapá	*Pellona* sp.	1	0.81
Aruanã	*Osteoglossum bicirrhossum*	2	0.68
Bacu	*Pterodoras granulosus*	1	0.01
Bico de Pato	*Hemisorubim platyrhynchos*	3	0.53
Bicudo	*Boulengerella ocellata*	4	0.90
Branquinha	*Potamorhina latior*	16	0.09
Cará	*Cichlasoma spectabile*	3	0.29
Caratinga	Not identified	1	0.04
Cascudinho	*Pareiorhapis duseni*	7	0.04
Cascudo	*Plecostomus* sp.	3	0.04
Coroatá	*Platynematichthys notatus*	2	1.86
Curimatã	*Prochilodus* sp.	38	0.12
Curimba	Not identified	1	0.08
Dourado	*Brachyplathystoma flavicans*	4	0.43
Filhote	*Brachyplathystoma filamentosum*	7	0.96
Jacundá	*Crenicichla reticulata*	5	0.37
Jaraqui	*Semaprochilodus taeniurus*	7	0.18
Jatuarana	*Hemiodus notatus*	9	0.10
Jaú	*Paulicea leutkeni*	2	0.52

Table 34. (Continued).

Popular name	Scientific name	N*	Hg (μg/g)
Lírio Braço de Moça	*Platystomatichthys sturio*	4	0.46
Mandi	*Pimelodus* sp.	6	0.34
Mandubé	*Ageneiosus brevifilis*	3	0.95
Mapará	*Hypophtalmus edentatus*	18	0.56
Matrinchã	*Brycon* sp.	1	0.05
Pacu	*Mylossoma* sp.	21	0.07
Peixe Cachorro	*Hydrolycus scomberoides*	9	0.97
Pescada	*Plagioscion squamosissimus*	14	0.33
Piau	*Laemiyta* sp.	4	0.07
Piau Apará	Not identified	3	0.21
Piau Aracu	*Laemolyta* sp.	6	0.16
Piau Botafogo	*Shizodon vittatum*	3	0.04
Piau Cabeça Gorda	*Leporinus friderici*	5	0.20
Piau Cabeça de Meia	*Schizodon fasciatum*	14	0.14
Pintado	*Pseudoplatystoma* sp.	22	0.92
Piramutaba	*Brachyplathystoma vaillanti*	1	0.09
Piranha	*Serrasalmus* sp.	46	0.80
Pirapitinga	*Colossoma bidens*	3	0.08
Pirarucu	*Arapaima gigas*	4	0.26
Sardinha	*Triportheus elongatus*	11	0.26
Surubim	*Pseudoplatystoma fasciatus*	1	0.38
Tambaqui	*Colossoma macropomum*	8	0.15
Traíra	*Hoplias malabaricus*	34	0.46
Tucunaré	*Cichla ocellaris*	49	0.36
Tucunaré açu	*Cichla* sp.	2	0.30

* N = number of samples.

Figure 27. Image of a specimen of *Platynematichthys notatus*. According to Coroatá (2009).

Table 35. Weight, length, and Hg content in fishes from the Madeira River basin, Rondônia State, Brazil. According to Gali (2002)

Number of samples	*Serrasalmus* sp. 46	*Prochilodus* sp. 38	*Hoplias malabaricus* 34	*Cichla ocellaris* 49
Weight (g)				
Minimum value	500	200	250	100
Maximum value	2100	2000	1200	2000
Average value	884	556	533	560
Length (cm)				
Minimum value	14	22	20.5	21
Maximum value	36.5	42.5	46	44
Average value	27	28	31	29
Hg (µg/g)				
Minimum value	0.1908	0.0029	0.1018	0.0627
Maximum value	2.1681	0.3380	1.1876	0.8432
Average value	0.8051	0.1203	0.4628	0.3551

Table 36. Statistical distribution of the mercury content in fishes from the Madeira River basin, Rondônia State, Brazil. According to Gali (2002)

Hg content range (µg/g)	Average Hg content (µg/g)	Frequency	Frequency percentage	Cumulative percentage
Serrasalmus sp. (N = 46)				
0.1908 - 0.2861	0.2400	4	8.7	8.7
0.2861 - 0.4289	0.3600	3	6.5	15.2
0.4289 - 0.6430	0.5400	12	26.1	41.3
0.6430 - 0.9641	0.8000	15	32.6	73.9
0.9641 - 1.4456	1.2000	8	17.4	91.3
1.4456 - 2.1675	1.8100	3	6.5	97.8
2.1675 - 3.2499	2.7100	1	2.2	100.0
Prochilodus sp. (N = 38)				
0.0018 - 0.0040	0.0029	1	2.6	2.6
0.0040 - 0.0088	0.0064	0	0	2.6
0.0088 - 0.0194	0.0141	0	0	2.6
0.0194 - 0.0429	0.0312	1	2.6	5.2
0.0429 - 0.0948	0.0689	8	21	26.2
0.0948 - 0.2095	0.1523	26	68.5	94.7
0.2095 - 0.4630	0.3366	2	5.3	100
Hoplias malabaricus (N = 34)				
0.1000 - 0.1506	0.1253	2	5.9	5.9
0.1506 - 0.2268	0.1887	2	5.9	11.8
0.2268 - 0.3416	0.2842	11	32.3	44.1
0.3416 - 0.5144	0.4280	9	26.5	70.6
0.5144 - 0.7747	0.6445	4	11.8	82.4
0.7747 - 1.1667	0.9707	5	14.7	97.1
1.1667 - 1.7570	1.4618	1	2.9	100
Cichla ocellaris (N = 49)				
0.0500 - 0.0771	0.0636	1	2	2
0.0771 - 0.1189	0.0980	1	2	4
0.1189 - 0.1833	0.1511	9	18.5	22.5
0.1833 - 0.2827	0.2331	14	28.6	51.1
0.2827 - 0.4359	0.3594	6	12.2	63.3
0.4359 - 0.6722	0.5542	15	30.6	93.9
0.6722 – 1.0366	0.8547	3	6.1	100

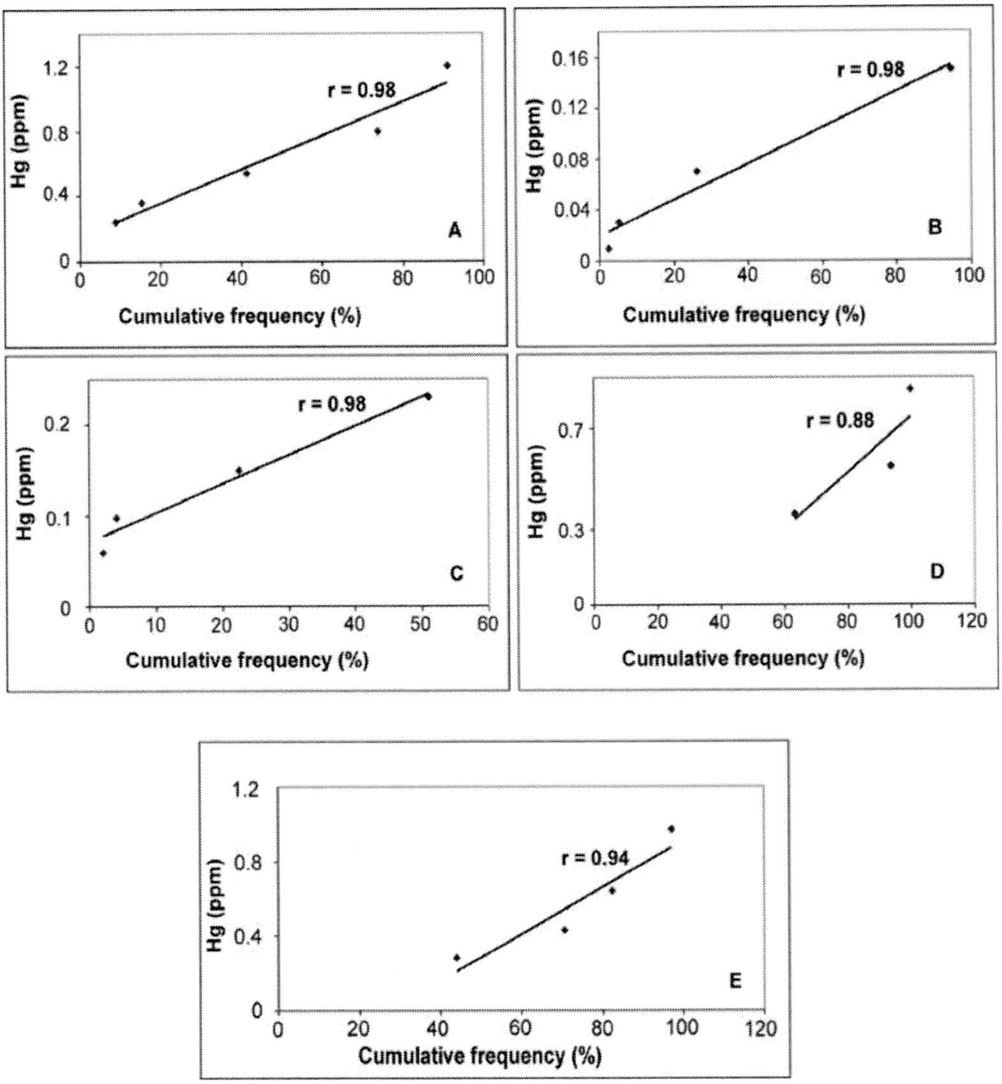

Figure 28. The average Hg content in a) *Serrasalmus* sp., b) *Prochilodus* sp., c) *Cichla ocellaris* (N=25), d) *Cichla ocellaris* (N=24), and e) *Hoplias malabaricus* plotted on a probability graph. According to Gali (2002).

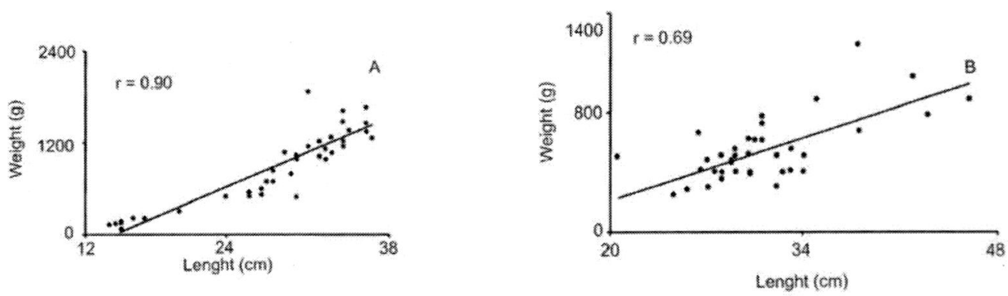

Figure 29. Continued on next page.

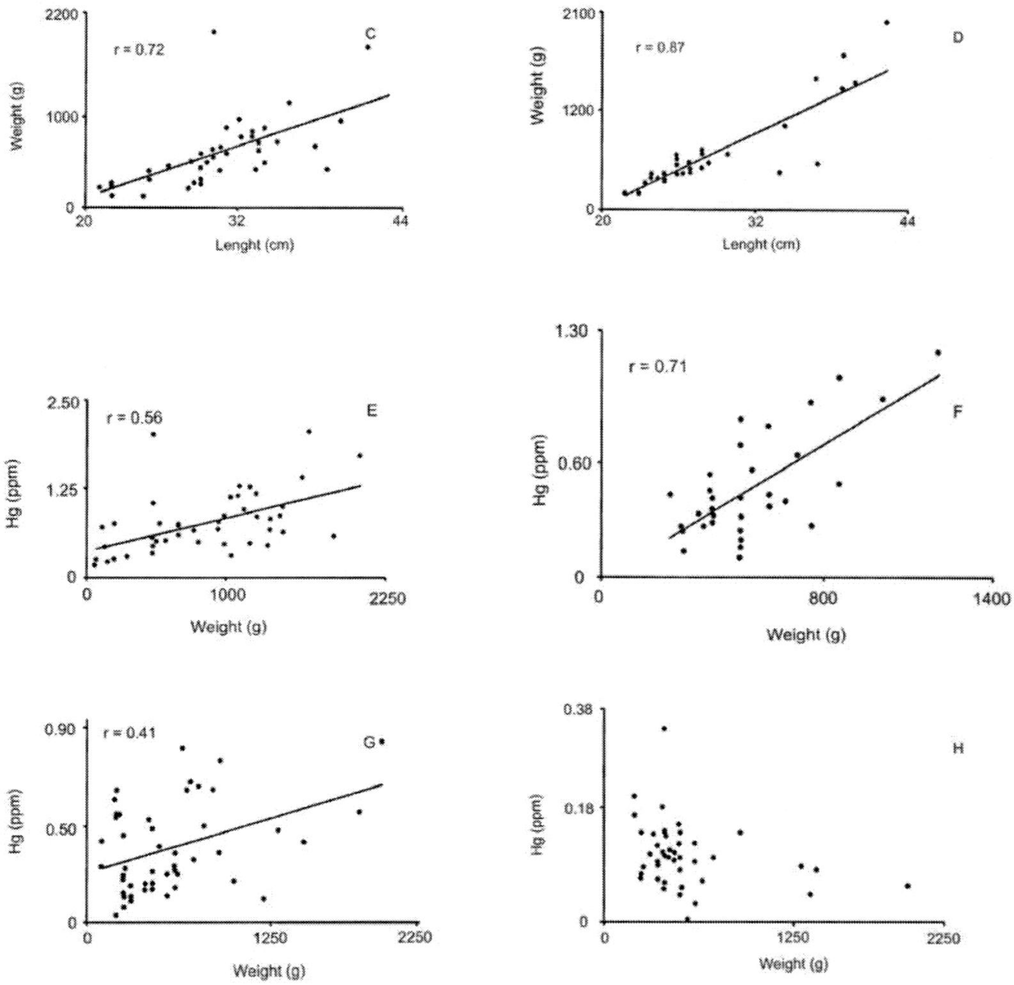

Figure 29. The relationship between weight and length in a) *Serrasalmus* sp., b) *Hoplias malabaricus*, c) *Cichla ocellaris*, d) *Prochilodus* sp., and the relationship between the Hg content and weight in e) *Serrasalmus* sp., f) *Hoplias malabaricus*, g) *Cichla ocellaris*, and h) *Prochilodus* sp. According to Gali (2002).

Chapter 12

The Future Challenges

The transitory and often clandestine small- and medium-scale mineral extraction operations named *garimpos* are ancient practices in the Amazon area, starting in the colonial times when the Europeans created legends and myths about the hidden richness in that vast region. The interest raised in 1542 when the Spanish sailor Francisco Orellana ventured along the Negro River, one of the Amazon Basin's great rivers, hunting a hidden city of gold (El Dorado) (Santos, 1981). Several mineral occurrences have been reported in the Amazon area since 1612 when it was firstly reported the presence of gold by *garimpeiros*, the persons working in *garimpos* (Santos, 1981). The total number of *garimpeiros* in the beginning of the 1970's has been estimated to range between 200,000 and 284,000 and despite the informality of the economic system, it had been a business responsible by billions of US dollars (Cleary, 1992).

The last gold rush in the Amazon area started in the late 1970's and 1980's at Serra Pelada (Hill of Gold), the very famous gold open pit mine in Pará State, north of Brazil, when thousands of men left their work in the fields of the brazilian north and northeast to join the gold mining activities. That epicenter favored the search for gold in different sources and regions of the Amazon area, mainly the alluvial gold that is found on the banks of rivers (Miller, 1997). As a consequence, many rivers and waterways in the Amazon have been exploited for gold using mercury in the mining process as an amalgam to separate the fine gold particles from other components in the bottom gravel (Malm et al., 1990). The alluvial method includes installation of equipment and housing, hydraulic pumping (high-pressure water to bring down the pebble embankment), concentration of gold by mercury, and burning the gold to remove the mercury (Eisler, 2006). During the peak activities involving the gold extraction, the metal was sold at specialized stores where it was again fired. These steps have been considered responsible for more than 70% of the mercury entering the environment. Mercury lost to rivers and soils as Hg° is comparatively unreactive, but the metallic mercury can undergo methylation in the river sediments, what contributes to its burden in fish and other biota and, consequently, to its entrance into the food chain (Eisler, 2006).

The elemental mercury discharged into the Amazon River basin due to gold mining activities in the late 1970's and 1980's involved mainly two tributaries of the Amazon River: the Madeira River and Tapajós River. Much attention has been given to mercury inputs at both rivers by numerous investigators as between 1987 and 1984 alone, more than 3000 metric tons of mercury was released into the biosphere of the Tapajós River basin (Villas Bôas, 1997).

The major populations at significant risk from mercury intoxication in Amazon gold mining communities consist on the riverine populations that routinely eat mercury-contaminated fish and gold dealers in indoor shops exposed to Hg° vapors (Lacerda and Salomons, 1992). Organic mercury levels have been identified among the Yanomama of the Brazilian Amazon basin by Sing et al. (2003). The Catrimani River basin in northern Brazil is the home of the Yanomama and has been the site of renegade gold mining since 1980. The inorganic mercury was released into the environment where it was organified and biomagnified in aquatic ecosystems. Ingestion of mercury-contaminated fish caused a potential hazard to fish-eating populations such as the Yanomama. Sing et al. (2003) surveyed Hg levels in Yanomama villagers living near mined and unmined rivers in 1994 and 1995, and analyzed Hg levels in piranha caught by villagers. In 1994, 90 Yanomama Indians from 5 villages and in 1995, 62 Yanomama Indians from 3 villages participated in their studies. Four villages surveyed in 1994 were located directly on the Catrimani River, approximately 140–160 km downstream from past gold-mining activities. The other village surveyed in 1994 was situated on the unmined Ajaraní River. In 1995, 2 of the Catrimani River villages were revisited, and a third Yanomama village, on the unmined Pacu River, was surveyed. Blood organic mercury levels among all villagers surveyed ranged from 0 to 62.6 μg/L (mean levels in each village between 21.2 μg/L and 43.1 μg/L). Mercury levels in piranha from the mined Catrimani River ranged from 235 to 1084 ng/g. Nine of 13 piranhas, measuring 30 cm or longer had total mercury levels which exceeded mercury consumption limits (500 ng/g) set by both the World Health Organization and the Brazilian Ministry of Health. Unexpectedly, high mercury levels were also observed in fish and villagers along the unmined Ajaraní and Pacu Rivers suggesting that indirect sources may contribute to environmental mercury contamination in the Amazon basin.

Gonçalves and Gonçalves (2004) presented additional important data generated by a multicentric study carried out by Brazilian researchers who, with the support of national and international institutions, worked during the 1990's to describe human exposure to mercury in gold mining areas in the Brazilian Amazon. Three sets of procedures were followed with residents of the Tocantins and Xingu River basins: (1) clinical examination, based on a standardized protocol and performed by a single researcher, in order to identify five basic levels of contamination; (2) determination of mercury levels in blood, hair, and urine, using atomic absorption spectrophotometry (Xingu Project); and (3) investigation of genotoxicity by using four cytogenetic indicators. During the first stage, 41 individuals were studied. Mercury was found in miners and their family members. Contamination was significantly associated with occupation and showed a differential correlation with genotoxicity indicators. After that, the Xingu project, which encompassed 625 individuals, focused on 417 Kayapó natives from the Gorotire and Djudjetiktire villages; 142 miners; and 66 riverine individuals (*ribeirinhos*). The highest levels of methylmercury in hair and total mercury in blood and urine were found among the Indian population. Contamination with inorganic forms of mercury was also observed in the blood, hair, and urine of Indians. Following a specific recommendation from the World Health Organization, special attention was given by Gonçalves and Gonçalves (2004) to pregnant Gorotire and Djudjetiktire women, who, although not directly exposed to mercury vapors, showed considerable levels of mercury intoxication.

The gold mining activities on the Madeira River extended from the Bolivian headwaters, on the Beni River, down to the city of Porto Velho, capital of Rondônia State. Because Porto Velho city is the most populated site (~370,000 inhabitants) along the Madeira River, a lot of investigations have been performed focusing the mercury presence in the Upper Madeira River, i.e. the drainage system upstream from Porto Velho city (figure 30). Downstream from those areas, riverside residents rely on fish consumption for subsistence and over the last 25 years several projects carried on the Madeira River basin produced a great amount of data on total Hg concentration in humans and different fish species. For instance, in July of 1993, hair samples were collected by Boischio and Cernichiari (1998) for mercury analysis from a group of mothers and their infants and one pregnant woman. By assuming a constant rate of hair growth (1.1 cm per month), a temporal profile of the methylmercury exposure was determined for the previous 2 to 3 years. The length of hair segments corresponded to hair growth during pregnancy and the subsequent breastfeeding periods. During all periods, hair mercury concentrations in six mothers were in the range of 4.5-26.8 μg/g, slightly lower than those of their infants (8.2-28.4 μg/g). Further segmental analyses of hair mercury from another six mothers showed concentrations in the range of 12.2-41.0 μg/g during the three trimesters of pregnancy and 4.0-33.5 μg/g during breast feeding-slightly lower than their infants (11.6-50.4 μg/g). Another four mothers showed hair mercury concentrations in the range of 21.3-84.4 μg/g, as reported Boischio and Cernichiari (1998), indicating the occurrence of mercury intoxication.

Figure 30. The Madeira River in the outskirts of Porto Velho city, Rondônia State, Brazil. According to Dias (2007).

The guideline reference values for total mercury recommended by the World Health Organization for fishes consumption is 0.5 μg/g for carnivorous and 0.3 μg/g for non-carnivorous ones (IPCS, 1990). In Brazil, the Hg limits established by the Ministry of Health correspond to 0.5 μg/g for non-carnivorous and 1.0 μg/g for carnivorous fishes (MS, 1998). Total mercury levels in fishes exceeding at least some of these tolerance limits have been reported many times in the Madeira River basin, as illustrated in table 34 for several specimens. Those data agree with some findings of Bastos et al. (2008) who measured the Hg concentration in 1,100 fishes from 86 species in the Madeira River basin that were collected between 1987 and 2000. Bastos et al. (2008) grouped the fishes according to their feeding habits, and the carnivorous ones were the best represented class with 34 species, from which 16 (8 commercially important) presented a mean total Hg value above the WHO threshold (0.5 μg/g); nine exclusively piscivorous species, despite low commercial interest, were all above the WHO threshold, showing the importance of the mercury biomagnifications, whereas among omnivorous fishes (20 species) four species exhibited mean total Hg content above of the WHO threshold and in three detritivorous species the WHO threshold (0.3 μg/g) was surpassed.

The previous investigation of Maurice-Bourgoin et al. (2000) also pointed out the mercury presence in two biotic compartments (fish and human hair) from the Upper Madeira River of the Bolivian Amazon basin. The results obtained from fish indicated, on one hand, that 86% of the piscivorous fishes collected in the Beni river were contaminated, and, on the other hand, that their high mercury concentrations could exceed by almost four times the WHO threshold safety limit. In the Beni River, the mercury concentrations found in omnivorous and mud-feeding fish ranged from 0.02 to 0.19 μg/g, and in piscivorous fish, from 0.33 to 2.30 μg/g. The mercury accumulated by carnivorous fishes was mainly present in its organic form; methylmercury represented 73-98% of the total mercury analyzed. Eighty persons were studied in the entire Bolivian Amazonian basin by Maurice-Bourgoin et al. (2000). Unlike the gold miners, who are more affected by tropical diseases, such as malaria and yellow fever, the indigenous people living on the banks of the Beni river, presented elevated levels of mercury (9.81 μg/g on average). Maurice-Bourgoin et al. (2000) also observed an increase in contamination in young children still being breast-fed, confirming that hair mercury concentration in babies was significantly affected by maternal mercury contamination during pregnancy. Their results showed that the major health impacts caused by mercury affect people who are not working directly in gold mining activities but who have a regular fish diet.

Despite table 32 only reports data for selenium in sediments from the Madeira River basin, others studies have claimed the importance of its evaluation in biotic compartments of that region. For instance, Dorea et al. (1998) analyzed selenium and mercury for samples of 7 species of piscivorous, omnivorous, and herbivorous fish caught at 12 different sites on the Madeira River basin. The mean selenium concentration ranged from 0.49 to 3.11 nmol/g and for mercury from 0.41 to 6.66 nmol/g depending on the fish species. The molar ratios of Hg:Se increased according to the fish trophic level. Piscivorous species had the highest mean ratio (4.0) and herbivorous species the lowest (0.9). There was a positive and statistically significant correlation between selenium and mercury concentrations for the herbivorous species ($r = 0.716$) not seen for omnivororus and piscivorous species ($r = -0.2032$). These findings were considered significant by Dorea et al. (1998) for the fish-eating population of the Madeira River because the ingestion of mercury would always be in excess of selenium.

The results of many site-specific investigations on the mercury concentrations in water, sediments, and biota of the Brazilian Amazon have indicated elevated values that are above the global averages, which have been often pointed out that they are a direct result of the widespread mercury amalgamation mining operations conducted by non-organized prospectors. However, geogenic rather than anthropogenic sources have also been considered relevant to explain the high Hg content values measured in abiotic and biotic compartments. In order to assess the regional impacts of Hg contamination from non-organized gold mining activities, Lechler et al. (2000) performed a systematic sampling of waters, sediments, and fishes in 1997 along a 900-km reach of the Madeira River. The sampling program extended from the Amazon River upstream to Porto Velho, the site of historic and ongoing mercury amalgamation mining. Mercury concentrations were found to be elevated above global averages in all sampled media. On the other hand, the geochemical data suggested that the high mercury levels are due largely to natural sources and natural biogeochemical processes, and that the impacts of anthropogenically released mercury from mine sites is relatively localized.

The lot of mercury concentration data on abiotic and biotic compartments here presented contributes to improve the knowledge about the Hg presence in the Brazilian Amazon, generating useful information for researchers/decision makers interested on its adequate management for diverse purposes. It has been shown the role of organic matter on the adsorption of mercury in lake sediments from the Madeira River basin, where Hg organic compounds are certainly being formed. This may be a subject of major concern as several specimens of fish ingest organic matter, Hg and microorganisms associated to mudstone at the bottom of lakes and margins of rivers. Thus, sediments act as important sinks for Hg in Amazon lakes and rivers, yielding important information relevant of the hydrological resources in the area, as may imply on the food-chain contamination. The data were acquired about 15-20 years after the nearly cessation of gold mining activities in the region, which reached its peak in the late 1980s. In spite of the reduction of Hg emissions to the Madeira River basin from gold mining proper, concentrations in fishes are practically similar to those measured during the gold rush, as also pointed out by Bastos et al. (2006). These authors also consider that the reduction in Hg concentrations is restricted to areas close to old point sources and only for abiotic compartments (air and sediments) and that the remobilization of Hg from bottom sediments plus re-emission from soils due to land use changes are probably responsible for keeping high Hg concentrations in biological samples.

Despite this scenario of anxiety, a number of social achievements in the 1990's have pointed to new standards of dignity in health care and of social ethics in Brazil that could benefit everyone, including the victims of environmental injury. However, the Madeira River is a major waterway in South America (~ 3,380 km long) and the longest tributary of the Amazon River, whose location is in the tri-border region of Peru, Bolivia and Brazil. Such exceptional conditions have favored the development of Integrated Regional Infrastructure for South America's (IIRSA) anchor projects (Amazon Watch, 2009). It would transform the Madre de Dios-Beni-Mamoré-Itenez-Madeira River system into a major corridor for energy production and raw material export. The proposal includes the construction of four hydroelectric dams, most importantly the Santo Antônio and Jirau dams in Rondônia State, Brazil. Together, these two dams would produce a projected 6,450 megawatts of hydroelectricity, totaling 8% of the Brazilian energy matrix (Amazon Watch, 2009). By comparison, this is equal to half of the electricity produced by Itaipu dam in the Brazilian state of Paraná, the second world's largest hydroelectric power plant. The Madeira project would also increase the capacity for transporting soybean, timber, and minerals to Pacific ports through the installation of navigation locks and dredging to open the river channel, and to connect with highways being built in the Peruvian and Bolivian Amazon (Amazon Watch, 2009).

The challenges for the establishment of the giant Madeira River Complex are enormous, involving cost, technical aspects and potential environmental and social impacts related to displacement, threats to food security, increased exposure to disease, pressure on social services, risks to biodiversity, and deforestation leading to greenhouse gas emissions (Amazon Watch, 2009). The Santo Antônio and Jirau dams would seriously harm the region's local populations and ecosystems, affecting four indigenous groups of the Upper Madeira River basin. Possible impacts created from the construction of the dams include the displacement of families and communities, the potential extinction of migratory fish species, the increased spread of malaria, increased erosion of river banks, threats of deforestation from agribusiness expansion, and stronger pressure on social services due to the migration of families in search of jobs (Amazon Watch, 2009).

Such worrisome scenario is at least as much as complex than that affecting the environmental questions associated to the mercury contamination in the Madeira River basin. Thus, it is presently hard to predict if all information already available for that area will be taken into account and properly utilized under a broader view of the decision makers, as it is well known that the social nature of the scientific activities implies that the data and information generated are intrinsically linked to social, economic and political influences and interests like it has been pointed out by the Science, Technology and Society (STS) relationships (Kneller, 1978).

The future will tell us which influences will be stronger…

REFERENCES

AAEC (Australian Atomic Energy Commission Research Establishment) Radionuclide migration around uranium ore bodies: analogue of radioactive waste repositories; Annual Report; AAEC: Sydney, 1983; 36 pp.

Adamy, A.; Romanini, S. J. In Programa de Levantamentos Geológicos Básicos do Brasil: Folhas Porto Velho, Mutumparaná, Jaciparaná e Abunã; DNPM (Departamento Nacional da Produção Mineral); Ed.; DNPM: Brasília, 1990.

Agaki, H.; Malm, O.; Kinjo, Y.; Harada, M.; Branches, F.; Pfeiffer, W. C.; Kato, H. In Proc. Int. Symp. Assessment of Environmental Pollution and Health Effects from Methylmercury; Kumamoto, 1993; p. 153.

Almeida, F. F. M. *Rev. Brasileira Geociências.* 1974, 4, 191-204.

Almeida, F. F. M. In Anais XX Congresso Brasileiro de Geologia; SBG (Sociedade Brasileira de Geologia); Ed.; SBG: Recife, 1978; v. 6, pp. 2393-2407.

Almeida, F. F. M. In Anais XXIII Congresso Brasileiro de Geologia; SBG (Sociedade Brasileira de Geologia); Ed.; SBG: Porto Alegre, 1984; v. 7, pp. 175-197.

Amaral, G. Geologia pré-cambriana da região Amazônica; Post PhD Thesis; USP: São Paulo, SP, 1974; 212 pp.

Amaral, G. In *O Pré-cambriano do Brasil*; Almeida, F. F. M.; Hasui, Y.; Eds.; Edgard Blücher: São Paulo, 1984; pp. 6-35.

Amazon Watch 2009 IIRSA Madeira River Complex [http://www.amazonwatch.org/amazon/BR/madeira/]

Andersson, A. In *The biogeochemistry of mercury in the environment*; Nriagu, J. O.; Ed.; Elsevier/North-Holland Biomedical Press: Amsterdam, 1979.

APHA (American Public Health Association) *Standard methods for the examination of water and wastewater*; APHA: Washington, DC, 1989; 17th edn.

Azcue, J. M.; Guimarães, J. R. D.; Mudroch, A.; Mudroch, P.; Malm, O. In *Handbook of techniques for aquatic sediments sampling*; Mudroch, A.; McKnight, S. D.; Eds.; CRC Press: Boca Raton, 1994; pp. 203-228.

Back, W.; Hanshaw, B. B. In *Advances in Hydroscience*; Chow, V.; Ed.; Academic Press: London, 1965; v. 2, pp. 49-109.

Bagnall, K. W. *The chemistry of selenium, tellurium and polonium*; Elsevier Publishing Co: Amsterdam, 1966.

Ball, J. W.; Nordstrom, D. K. WATEQ4F: a program for the calculating speciation of major, trace, and redox elements in natural waters; Colorado School of Mines: Golden, USA, 1992.

Barbosa, G. V.; Pinto, M. N. In Projeto RADAM: Folha SA-23 São Luís e parte da Folha SA-24 Fortaleza; DNPM (Departamento Nacional da Produção Mineral); Ed.; DNPM: Rio de Janeiro, 1973; v. 3.

Barner, H. L.; Rosemberger, B. *Geol. Soc. Am. Sp. Paper.* 1963, 73, 1-110.

Bastos, W. R.; Rebelo, M. de F.; Fonseca, M. de F.; Almeida, R. de; Malm, O. *Acta Amaz.* 2008, 38 (3), 431-438.

Bastos, W. R.; Gomes, J. P. O.; Oliveira, R. C.; Nascimento, E. L.; Bernardi, J. V. E.; Lacerda, L. D.; Silveira, E. G.; Pfeiffer, W. C. *Sci. Total Environ.* 2006, 368 (1), 344-351.

Bellízia, C. M. *Bol. Geol.* 1972, 6, 251 - 305.

Benes, P. In *The behavior of radium in waterways and aquifers*; IAEA (International Atomic Energy Agency); Ed.; IAEA: Vienna, 1984; pp. 118-173.

Benes, P.; Havlík, B. In *The biogeochemistry of mercury in the environment*; Nriagu, J. O.; Ed.; Elsevier/North-Holland Biomedical Press: Amsterdam, 1979.

Bethke, P. J. *Bull. Geol. Soc. Am.* 1956, 67, 1671.

Bettencourt, J. S.; Dall'Agnol, R. In Excursion Guide Int. Symp. Granites and Associated Mineralization; Sup. Geol. Rec. Minerais; Ed.; SME: Salvador, 1987; pp. 49-87.

Bettencourt, J. S.; Tosdal, R. M.; Leite Jr., W. B.; Payolla, B. L. In Excursion Guide Symp. Rapakivi Granites and Related Rocks; UFPA; Ed.; UFPA: Belém, 1995; pp. 5-6.

Björnberg, A.; Hakason, L.; Lundberg, K. *Environ. Pollut.* 1988, 49, 53-61.

Boaventura, R. G. Sistema de reação em série para determinação de mercúrio com geração de vapor; PhD Thesis; Universidade de Brasília: Brasília, 1995; 143 pp.

Boischio, A. A.; Cernichiari, E. *Environ. Res.* 1998, 77 (2), 79-83.

Bonotto, D. M. *J. South American Earth Sci.* 1998, 11, 389-405.

Bonotto, D. M.; Mancini, L. H. *Geochim. Brasil.* 1992, 6, 153-167.

Bonotto, D. M.; Andrews, J. N. *Chem. Geol.* (*Isot Geosci Sec*) 1993, 103, 193-206.

Bonotto, D. M.; Andrews, J. N. *Appl. Radiat. Isot.* 2000, 52, 965-983.

Bonotto, D. M.; Andrews, J. N.; Darbyshire, D. P. F. *Appl. Radiat. Isot.* 2001, 54, 977-994.

Bonotto, D. M.; Vergotti, M.; Silveira, E. G. In *Sediment Dynamics and the Hydromorphology of Fluvial Systems*; Rowan, J. S.; Duck, R. W.; Werritty, A.; Eds.; IAHS: Wallingford, 2006; v. 1, pp. 355-362.

Borovec, Z. *Chem. Geol.* 1981, 32, 45-58.

Branco, S. M. *O desafio amazônico*; Moderna: São Paulo, 1989.

Brandini, F. 2005 Mar dulce mar. Amigos da terra – Amazônia Brasileira [http://www.amazonia.org.br/ opiniao/artigo_detail.cfm?id=166380]

Caprioglio, L.; Bonotto, D. M. In Resumos IX Congresso Iniciação Científica; UNESP; Ed.; UNESP: Ilha Solteira, 1997; p. 91.

Carrol, D. *GSA Bulletin. 1*959, 70, 749-780.

CETEM (Centro de Tecnologia Mineral) Estudos biogeoquímicos sobre o mercúrio em ambientes aquáticos de Poconé; *Tech. Rep.* 12/91; CETEM: Rio de Janeiro, 1991a; 90 pp.

CETEM (Centro de Tecnologia Mineral) Poconé: um campo de estudos do impacto ambiental do garimpo; *Environmental Technology Series;* CETEM: Rio de Janeiro, 1991b; 113 pp.

Cleary, D. *A garimpagem de ouro na Amazônia: uma abordagem antropológica*; UFRJ: Rio de Janeiro, 1992; 237 pp.

Cleary, D.; Thornton, I. L.; Brown, N.; Kazantzis, G.; Delves, T.; Worthington, S. *Nature.* 1994, 369, 613-614.

Colman, S. M. *US Geol. Surv. Prof. Paper* 1982, 1245, 1-51.

Cordani, U. G. In *Uranium deposits in Latin America: geology and exploration*; IAEA (International Atomic Energy Agency); Ed.; IAEA: Vienna, 1981; pp. 3-23.

Cordani, U. G.; Tassinari, C. C. G.; Teixeira, W.; Basei, M. A.; Kawashita, K. In Actas II Congr. Geol. Chileno; Arica 1979; v. 4; p. 137-148.

Coroatá 2009 Platynematichthys notatus [http://www.fishbase.org.cn/Summary/SpeciesSummary.php?id=121238&lang=Portuguese_po]

Costa, J. B. S.; Hasui, Y. In III Simp. Nacional Estudos Tectônicos; SBG (Sociedade Brasileira de Geologia); Ed.; SBG: Rio Claro, 1991; v. 1, pp. 141-145.

Costa Neto, A.; Bonotto, D. M.; Mesquita, F. L. A. In *Environmental Health Risk II*; Brebbia, C. A.; Fayzieva, D.; Eds.; WIT Press: Southampton, 2003; pp. 3-14.

Cotton, A. F.; Wilkinson, G. *Química inorgânica*; Livros Técnicos e Científicos Editora S.A.: Rio de Janeiro, 1982.

Cramer, S. W. *Int. Min.* 1990, 29-30.

Cutter, G. A. *Anal. Chem.* 1985, 57, 2951-2955.

Dana, J. D. *Manual de Mineralogia*; Livros Técnicos e Científicos Editora S.A.: Rio de Janeiro, 1976; 645 pp.

Dias, W. 2007 Rio Madeira [http://en.wikipedia.org/wiki/File:Rio_Madeira_ 09072007.jpg]

Dickson, F. W.; Tunell, G. *Am. J. Sci.* 1958, 256, 654-679.

D'Itri, P. A.; D'Itri, F. A. *Mercury contamination: a human tragedy*; John Wiley: New York, 1977; 85 pp.

DNPM (Departamento Nacional da Produção Mineral) Geologia, geomorfologia, pedologia, vegetação e uso da terra; DNPM-Projeto RADAM: Rio de Janeiro, 1978.

Dolbec, J.; Mergler, D.; Passos, C. J. S.; Morais, S. S.; Lebel, J. *Int. Arch. Occup. Environ. Health.* 2000, 73, 195-203.

Dolbec, J.; Mergler, D.; Larribe, F.; Roulet, M.; Lebel, J.; Lucotte, M. *Sci. Total. Environ.* 2001, 271, 87-97.

Domínguez, C. In *Issues of local and global use of water from the Amazon*; Aragón, L.E.; Clusener-Godt, M.; Eds.; UNESCO: Montevideo, 2004; pp. 77-100.

Eisler, R., *Mercury hazards to living organisms*; CRC Press: Potomac, Maryland, 2006; 336 pp.

ELETRONORTE (Centrais Elétricas do Norte do Brasil S/A) Geologia, condições de fundação e materiais naturais de construção dos sítios Santo Antônio, Teotônio e Girau, no Rio Madeira; Preliminary Report; ELETRONORTE: Rio de Janeiro, 1987; 85 pp.

Erhart, H. *Itinéraires geochimiques et cycle géologique de l`aluminium*; Doin: Paris, 1973.

Evans, J. W. *Quart. J. Geol. Soc. London.* 1906, 62, 88-124.

Faure, G. *Principles and applications of inorganic geochemistry*; Macmillan Publishing Co.: New York, 1991; 626 pp.

FIBGE (Fundação Instituto Brasileiro de Geografia e Estatística) Geografia do Brasil: Região Norte; IBGE: Rio de Janeiro, 1991; v. 3, 466 pp.

Fisher, R. A.; Yates, F. *Statistical tables*; Oliver & Boyd: Edinburgh, 1953; p. 40.

Förstner, U.; Salomons, W. *Environ. Technol. Letts.* 1980, 1, 494-505.

Frenet, M. Phenomenes de fixation - desorption du mercure sur les argiles dans les eaux a salinité variable. Aplication a l'estuaire de La Loire; These (Docteur en Sciences Naturelles); Universite de Nantes: Nantes, 1979; 227 pp.

Gali, P. A. S. O Hg em peixes de corpos d´água da porção norte do Estado de Rondônia; Ms Dissertation; UNESP: Rio Claro, SP, 2002; 76 pp.

Gambrel, R. P.; Khalid, R. A.; Patrick Jr., W. H. *Environ. Sci. Technol.* 1980, 14, 431-436.

García, C.; Cabrera, D.; Mejía, L.; Rojas, A. *Suelos Ecuatoriales.* 1999, 29 (2), 150-154.

Garrels, R. M.; Christ, C. L. *Solutions, minerals and equilibria*; Harper & Row: New York, 1965; 450 pp.

Gaudette, H. E.; Olszewski Jr, W. *J. Geol. Mijnb.* 1985, 64, 131-144.

Gibbs, R. J. *Science.* 1970, 170, 1088-1090.

Gibbs, A.; Barron, C. N. *Episodes.* 1983, 6, 7-14.

Gonçalves, A.; Gonçalves, N. N. S. *PAJPH.* 2004, 16 (6), 415-419.

Goulding, M. Ecologia da pesca do rio Madeira; INPA-Instituto Nacional de Pesquisas Amazônicas: Manaus, 1979; 172 pp.

Greenwood, N. N.; Earnshaw, A. *Chemistry of the elements*; Pergamon Press: Frankfurt, 1991; 412 pp.

Guimarães, J. R. D.; Roulet, M.; Lucotte, M.; Mergler, D. *Sci. Total Environ.* 2000a, 261, 91-98.

Guimarães, J. R. D.; Meili, M.; Hylander, L. D.; Castro e Silva, E.; Roulet, M.; Mauro, J. B. N.; Lemos, R. A. *Sci. Total Environ.* 2000b, 261, 99-107.

Hach *Water analysis handbook*; Hach Co.: Loveland, 1992; 2nd edn., 831 pp.

Hacon, S.; Rochedo, E. R. R.; Campos, R. R. R.; Lacerda, L. D. *J. Geochem. Explor.* 1997, 58, 209-216.

Hacon, S.; Artaxo, P.; Campos, R. C.; Cont, L. F.; Lacerda, L. D. *Water Air Soil. Pollut.* 1995, 80, 273-283.

Harriss, R. C. *Science*. 1978, 170, 736-737.

Hasui, H.; Almeida, F. F. M. *Episodes.* 1985, 8, 29-37.

Hasui, H.; Haralyi, N. I. E.; Schobbenhaus, C. In Anais II Simp. Geol. Amazônia; DNPM (Departamento Nacional da Produção Mineral); Ed.; DNPM: Manaus, 1984; pp. 129-147.

Hem, J. D. *U.S. Geol. Surv. Prof. Paper.* 1970, 713, 19-28.

Hosobara, K.; Kozuma, H.; Kawasaki, K.; Tsuruta, T. *Nippon Kagaku Zasshi.* 1961, 82, 1479-1480.

IBGE (Instituto Brasileiro de Geografia e Estatística) Mapa de vegetação do Brasil; MEFP-Diretoria de Geociências: Brasília, 1993.

IBGE (Instituto Brasileiro de Geografia e Estatística) 2007 O censo do Brasil [http://www.sidra.ibge.gov.br]

IPCS (International Program on Chemical Safety) *Environmental health criteria.* 101: methyl-mercury; WHO: Geneva, 1990; 141 pp.

IPCS (International Program on Chemical Safety) *Environmental health criteria: inorganic mercury*; UNESCO: Geneva, 1991; v. 118, p. 162.

Isotta, C. A. L.; Carneiro, J. M.; Kato, H. T.; Barros, R. J. Projeto província estanífera de Rondônia; DNPM/CPRM: Porto Velho, 1978; v. 1-3.

Jonasson, I.; Boyle, R. W. 1979. In Effects of mercury in the Canadian environment; NRC (National Research Council of Canada); Ed.; NRC: Ottawa, 1979; pp. 28-49.

Jovanovic, S.; Reed, G. W. *Geochim. Cosmochim. Acta.* 1968, 32, 341-346.

Kabata-Pendias, A.; Pendias, H. *Trace elements in soils and plants*; CRC Press Inc.: Boca Raton, 1984.

Kharaka, Y. K.; Ambats, G.; Presser, T. S. *Appl. Geochem.* 1996, 11, 979-802.

Kiehl, E. J. Interpretation of soil properties; Teaching text; ESALQ (University of Agriculture "Luiz de Queirós", USP-University of São Paulo: Piracicaba, SP, 1977; 237 pp.

Kneller, G. F. *Science as a human endeavor*; Columbia University Press: New York, 1978; 333 pp.

Krauskopf, K. B. *Introdução à geoquímica*; Editora Polígono: São Paulo, 1972; v. 1-2, 295 pp.

Krauskopf, K. B. *Introduction to geochemistry;* McGraw-Hill: New York, 1979; 617 pp.

Krauskopf, K. B.; Bird, D. K. *Introduction to geochemistry;* Mc Graw-Hill: New York, 1995; 546 pp.

Lacerda, L. D. *Nature*. 1995, 374, 21-22.

Lacerda, L. D.; Salomons, W. Mercúrio na Amazônia: uma bomba relógio química; CETEM: Rio de Janeiro, 1992; *Série Tecnologia Ambiental.* 3, 78 pp.

Lacerda, L. D.; Marins, R. V. *J. Geochem. Explor.* 1997, 58, 223-229.

Lacerda, L. D.; Pfeiffer, W. C.; Ott, A. T.; Silveira, E. G. *Biotropica.* 1989, 21, 91-93.

Lacerda, L. D.; Pfeiffer, W. C.; Silveira, E. G.; Bastos, W. R.; Souza, C. M. M. In Anais I Congr. Bras. Geoquímica; SBGq (Sociedade Brasileira de Geoquímica); Ed.; SBGq: Porto Alegre, 1987; pp. 295-299.

Lacerda, L. D.; Malm, O.; Guimarães, J. D. R.; Salomons, W.; Wilken, R. D. In *Biogeodynamics of pollutants in soils and sediments*; Salomons, W.; Stigliani, W.; Eds.; Springer-Verlag: Berlin, 1995; pp. 213-245.

Lacerda, L. D.; Pfeiffer, W. C.; Marins, R. V.; Rodrigues Filho, S.; Souza, C. M. M.; Bastos, W. R. *Water Air Soil Pollut.* 1991, 55, 283-294.

Langmuir, D. *Geochim. Cosmochim. Acta.* 1978, 42, 547-569.

Leal, J. W.; Silva, H.; Santos, D. B.; Teixeira, W.; Lima, M. I. C.; Fernandes, C. A. C.; Pinto, A. C. In Projeto RADAM: Folha NA/NB 22 Macapá; DNPM (Departamento Nacional da Produção Mineral); Ed.; DNPM: Rio de Janeiro, 1978; v. 6.

Lechler, P. J.; Miller, J. R. In Proc. Int. Symp. Perspectives for Environmental Geochemistry in Tropical Countries; Abrão, J. J.; Wasserman, J. C.; Silva-Filho, E. V.; Eds.; SBGq: *Niterói,* 1993; pp. 433-436.

Lechler, P. J.; Miller, J. R.; Lacerda, L. D.; Vinson, D.; Bonzongo, J. C.; Lyons, W. B.; Warwick, J. J. *Sci. Total. Environ.* 2000, 260 (1-3): 87-96.

Lee, J. D. *Fundamentos da química inorgânica*; Edgard Blücher: São Paulo, 1971; 249 pp.

Lee, J. D. *Química inorgânica não tão concisa*; Edgard Blücher: São Paulo, 1996; 331 pp.

Lima, M. I. C.; Montalvão, R. M. G.; Issler, R. S.; Oliveira, A. S.; Basei, M. A. S.; Araújo, J. F. V.; Silva, G. G. In Projeto RADAM: Folha NA/NB 22 Macapá; DNPM (Departamento Nacional da Produção Mineral); Ed.; DNPM: Rio de Janeiro, 1978; v. 6.

Lindqvist, O.; Jernelov, A.; Johansson, K.; Rodhe, H. Global and local sources; National Swedish Protection Board: Sölna, 1984; Rep. PM 1816.

Lisboa, P. L. B. Rondônia - Colonização e Floresta; Programa Polonoroeste, Tech. Rep. 9; CNPq: Brasília, 1989; 216 pp.

Litherland, M.; Klinck, B. A.; O'Connor, E. A.; Pitfield, P. E. *J. Nature*. 1985, 314, 345-348.

Litherland, M.; Annells, R. N.; Darbyshire, D. P. F.; Fletcher, C. J. N.; Hawkins, M. P.; Klinck, B. A.; Mitchell, W. I.; O'Connor, E. A.; Pitfield, P. E. J.; Power, G.; Webb, B. C. *Precambrian Res.* 1989, 43, 157–174.

Luoma, S. N. *Sci. Total Environ.* 1983, 28, 1-22.

Lobato, F. P. N.; Appel, L. E.; Godoy, M. C. F. T.; Ritter, J. E. *Bol. Div. Fom. Prod. Min.* 1966, 125, 1-209.

Lockwood, R. A.; Chen, K. Y. *Environ. Sci. Technol.* 1973, 7, 1028-1034.

Loring, D. H. *Can. J. Earth Sci.* 1975, 12, 1219-1237.

Malm, O. Contaminação ambiental e humana por mercúrio na região garimpeira de ouro do rio Madeira. PhD Thesis; UFRJ: Rio de Janeiro, 1991; 113 pp.

Malm, O.; Pfeiffer, W. C.; Souza, C. M. M. In Proc. VIII Int. Conf. Heavy Metals in the Environment; Edinburgh, 1991; v. 1, pp. 515-518.

Malm, O.; Guimarães, J. R. D.; Pfeiffer, W. C. In Proc. Int. Symp. Perspectives for Environmental Geochemistry in Tropical Countries; Abrão, J. J.; Wasserman, J. C.; Silva-Filho, E. V.; Eds.; SBGq: Niterói, 1993; pp. 391-393.

Malm, O.; Pfeiffer, W. C.; Bastos, W. R.; Souza, C. M. M. *Cien. Cult.* 1989, 41, 88-92.

Malm, O.; Pfeiffer, W. C.; Rodrigues, W. B.; Magalhães, C. M. S.; Rudolf, R. *Ambio.* 1990, 19, 11-15.

Manasse, E. *Mem. Soc. Sci. Nat.* 1921, 156-163.

Martinelli, L. A.; Ferreira, J. R.; Forsberg, B. R.; Victoria, R. L. *Ambio.* 1988, 17, 252-254.

Mason, B. H. *Principles of geochemistry;* Wiley: New York, 1966; 329 pp.

Mason, B. H. *Princípios de geoquímica;* Polígono: São Paulo, 1971; 403 pp.

Maurice-Bourgoin, L.; Quiroga, I.; Guyot, J. L.; Malm, O. *Ambio.* 1999, 28, 302-306.

Maurice-Bourgoin, L.; Quiroga, I.; Chincheros, J.; Courau, P. *Sci. Total Environ.* 2000, 260, 73-86.

Maurice-Bourgoin, L.; Guyot, J. L.; Seyler, P.; Quintanilla, J.; Courau, P. IAHS Publication 1997, 243, 85-92.

McNeal, J. M.; Rose, A. W. *Geochim. Cosmochim. Acta.* 1974, 38, 1759-1784.

McNeal, J. M.; Feder, G. L.; Wilber, W. G.; Deverel, S. J. In Proc. U.S. Geological Survey Workshop on Environmental Geochemistry; Doe, B. R.; Ed.; U.S. Geological Survey: Denver, 1990; pp. 177-178.

Melo, D. P.; Costa, R. C. R.; Natali Filho, T. In Projeto RADAM: Folha SC.20 Porto Velho; DNPM (Departamento Nacional da Produção Mineral); Ed.; DNPM: Rio de Janeiro, 1978.

Miller, J. R. *J. Geochem. Explor.* 1997, 58, 101-118.

Mitra, S. Mercury in the ecosystem: its dispersion and pollution today; Trans. Tech. Pub. Ltd.: Switzerland, 1986; 312 pp.

Moore, J. W.; Ramamoorthy, S. *Heavy metals in natural waters*; Springer-Verlag: New York, 1984; 210 pp.

MS (Ministério da Saúde) Princípios gerais para o estabelecimento de níveis máximos de tolerância para contaminantes inorgânicos: limites máximos de tolerância para contaminantes inorgânicos; Rule No. 685, DOU 08/28/1998; ANVISA-Agência Nacional de Vigilância Sanitária: Brasília, 1998.

Nehmi, V. *Química*; Editora Ática: São Paulo, 1995; 496 pp.

Nriagu, J.O. In Proc. Int. Symp. Perspectives for Environmental Geochemistry in Tropical Countries; Abrão, J. J.; Wasserman, J. C.; Silva-Filho, E. V.; Eds.; SBGq: *Niterói,* 1993; pp. 365-368.

Nriagu, J. O.; Pfeiffer. W. C.; Malm, O.; Souza, C. M. M.; Mierle, G. *Nature*. 1992, 356, 389.

Nunes, D. D. Hidrovia do Rio Madeira; Research Rep. Pibic-Unir/Cnpq; UNIR-Fundação Universidade Federal de Rondônia: Porto Velho, 1999.

Payolla, B. L. As rochas graníticas e sieníticas das cachoeiras Teotônio e Santo Antônio, rio Madeira, Porto Velho, Rondônia: geologia, petrologia e geoquímica; Ms Dissertation; Universidade de Brasília: Brasília, 1994; 145 pp.

Peazê, L. 2005 Contaminando a vida [http://www.luispeaze.com/minamata.htm].

Pedro, G. *Rev. Geogr. Phys. Géol. Dyn.* 1968, X (5), 457-470.

Pedro, G.; Sieffermann, G. *Review in modern problems of geochemistry*; UNESCO: Paris, 1979; p. 39-55.

Pfeiffer, W. C.; Lacerda, L. D. *Environ. Technol. Lett.* 1988, 9, 325-350.

Pfeiffer, W. C.; Lacerda, L. D.; Malm, O.; Souza,C. M. M.; Silveira, E. G.; Bastos, W. R. *Sci. Total. Environ.* 1989, 87/88, 233-240.

Pfeiffer, W. C.; Lacerda, L. D.; Malm, O.; Souza, C. M. M.; Lacerda, L. D.; Silveira, E. G. *Ciência Hoje*. 1990, 1, 13-19.

Pfeiffer, W. C.; Lacerda, L. D.; Malm, O.; Souza, C. M. M.; Silveira, E. G.; Bastos, W. R. *Forest Ecol. Manag.* 1991, 38, 239-245.

Piper, A. M. *Trans. Amer. Geophysical. Union*. 1944, 25, 914-928.

Ramamoorthy, S.; Rust, B. R. *Can. J. Earth Sci.* 1976, 13, 530-536.

Ramamoorthy, S.; Rust, B. R. *Environ. Geol.* 1978, 2, 165-172.

Ribeiro, J. S.; Santos, U. *Acta Amaz.* 1983, 1, 84-93.

Richardson, S. M.; McSween Jr., H. Y. *Geochemistry: pathways and processes*; Prentice-Hall: New Jersey, 1989; 487 pp.

Rodrigues Filho, S.; Maddock, J. E. L. *J. Geochem. Explor.* 1997, 58, 231-240.

Rosenfeld, I.; Beath, O. *Selenium geobotany, biochemistry, toxicity and nutrition*; Academic Press: New York, 1964.

Roulet, M.; Lucotte, M.; Guimarães, J. R. D.; Rheault, I. *Sci. Total Environ.* 2000a, 261, 43-59.

Roulet, M.; Lucotte, M.; Canuel, R.; Farella, N.; Courcelles, M.; Guimarães, J. R. D.; Mergler, D.; Amorim, M. *Chem. Geol.* 2000b, 165, 243-266.

Roulet, M.; Lucotte, M.; Canuel, R.; Rheault, I.; Tran, S.; Freitos Gog, Y. G.; Farella, N.; Vale, R. S.; Passos, C. J. S.; Jesus da Silva, E.; Mergler, D.; Amorim, M. *Sci. Total Environ.* 1998, 213, 203-211.

Salomons, W.; Förstner, U. *Metals in the hydrocycle*; Springer-Verlag: Berlin, 1984; 353 pp.

Salomons, W.; Stigliani, W. M. *Biogeodynamics of pollutants in soils and sediments*; Springer: Heidelberg, 1995; 352 pp.

Santos, B. A. *Amazônia: potencial mineral e perspectivas de desenvolvimento*; EDUSP: São Paulo, 1981; 256 pp.

Santos, J. O. In *Geologia do Brasil*; Schobbenhaus, C.; Campos, D. A.; Derze, F. R.; Asmur, H. E.; Eds.; DNPM (Departamento Nacional da Produção Mineral): Brasília, 1984; pp. 57-88.

Santos, G. M. Pesca e ecologia dos peixes de Rondônia; PhD Thesis; INPA: Manaus, 1991; 213 pp.

Santos, J. O.; Loguércio, S. O. C. In *Geologia do Brasil*; Schobbenhaus, C.; Campos, D. A.; Derze, F. R.; Asmur, H. E.; Eds.; DNPM (Departamento Nacional da Produção Mineral): Brasília, 1984; pp. 93-127.

Santos, M. U.; Ribeiro, M. N. *Acta Amaz.* 1988, 18, 145-172.

Santos, M. U.; Ribeiro, M. N.; Tancredi, C. F. N. S. *Acta Amaz.* 1987, Sp. No., 143-150.

Santos, A.; Ribeiro, M. N. G.; Ribeiro, J. S.; Bringel, S. R. B. *Acta Amaz.* 1981, 11, 335-346.

Saukov, A. A. Akademii Nauk SSSR 1946, 78, 129.

Scandolara, J. E.; Rizotto, G. J.; Silva, C. R.; Babia, R. B. C.; Quadros, M. L. E. S.; Amorim, J. L. Mapa geológico de Rondônia e áreas adjacentes; CPRM-Companhia de Pesquisa e Recursos Minerais: Porto Velho, 1996.

Schellmann, W., In Proc. I Int. Symp. Laterization Processes; Trivandrum, 1980; pp. 1-10.

Schimidt, G. W. *Amazoniana.* 1972, 3, 199-207.

Schwarzenbach, G.; Wildmer, M. *Helvet. Chim. Acta.* 1963, 46, 2613-2628.

Shcheglov, A. D. *Int. Geol. Rev.* 1970, 12, 1453-1497.

Siegel, S. M.; Siegel, B. Z. *Environ. Sci. Technol.* 1975, 9, 473-474.

Silva, L. G. T. Estudo pedogeoquímico das coberturas de alteração intempérica que ocorrem na região de Marabá-PA: uma contribuição ao manejo ambiental; Ms Dissertation; UNESP: Rio Claro, SP, 1995.

Silva, A. P. Mercúrio em ambientes aquáticos de Poconé/MT; PhD Thesis; USP: São Paulo, 1996; 118 pp.

Silveira, E. G. Mobilização do mercúrio e outros elementos no Rio Madeira (RO) entre as cachoeiras de Teotônio e Santo Antônio; PhD Thesis; UNESP: Rio Claro, SP, 1998; 93 pp.

Sing, K. A.; Hryhorczuk, D.; Saffirio, G.; Sinks, T.; Paschal, D. C.; Sorensen, J.; Chen, E. H. *Ambio.* 2003, 32 (7), 434-439.

Souza, E. C.; Marques, W. J. Granito porfiroblástico de Santo Antônio, Território Federal de Rondônia, Brasil; CPRM-Companhia de Pesquisa de Recursos Minerais: Porto Velho, 1974; 11 pp.

Summerfield, M. A. *Global geomorphology: an introduction to the study of landforms*; Longman & Wiley: New York, 1994; 538 pp.

Szalay, A. *Geochim. Cosmochim. Acta.* 1964, 28, 1605-1614.

Tancredi, A. C. F. S.; Lobato, J. R. B.; Serra, V. H. *Acta Amaz.* 1987, 16/17, 293-316.

Tassinari, C. C. G. A evolução geotectônica da província Rio-Juruena na região Amazônica; Ms Dissertation; USP: São Paulo, 1981.

Teixeira, W.; Tassinari, C. C. G. In Anais II Simp. Geol. Amazônia; DNPM (Departamento Nacional da Produção Mineral); Ed.; DNPM: Manaus, 1984; pp. 87-91.

Teixeira, W.; Tassinari, C. C. G.; Cordani, U. G.; Kawashita, K. *Precambrian Res.* 1989, 42, 213-227.

Todd, D. K. *Hidrologia de águas subterrâneas*; Editora Edgard Blücher: São Paulo, 1959; 170 pp.

Urenha Jr., A. 2008 Peixes do Brasil [http://pesca-cia.uol.com.br/peixes-do-brasil/agua-doce.aspx?c=1695]

USBM (U.S. Bureau of Mines) *Mineral facts and problems.* U.S. Department of the Interior: Washington, 1970.

Veiga, M. M.; Fernandes, F. R. C.; Farid, L. H.; Machado, J. E. B.; Silva, A. O.; Lacerda, L. D.; Silva, A. P.; Silva, E. C.; Oliveira, E. F.; Silva, G. D.; Pádua, H. B.; Pedrozo, L. R.

M.; Ferreira, N. L. S.; Ozaki, S. K.; Marins, R. V.; Imbassahy, J. A.; Pfeiffer, W. C.; Bastos, W. R.; Souza, V. P. Poconé: um campo de estudos do impacto ambiental do garimpo. CETEM/CNPq: Rio de Janeiro, 1991; STA-Série Tecnologia Ambiental, 1, 113 pp.

Velinsky, D. J.; Cutter, G. A. 1991 *Geochim. Cosmochim. Acta.* 1991, 55, 179-191.

Vergotti, M. O mercúrio em lagos da bacia do Rio Madeira (RO); Ms Dissertation; UNESP: Rio Claro, SP, 2002.

Verschure, R. H.; Bon, E .H. Geology and geochronology of tin bearing volcanic-plutonic complexes in Rondônia (western Brazil): refletion on the genesis of this type of magmatism; Research Rep. 1970/72; Z.W.O. Labor, Voo Isotopen-Geology: Amsterdam, 1972; pp. 177-198.

Villas Bôas, R. C. *J. Geochem. Explor.* 1997, 58, 217-222.

Vinogradov, A. P. *Geochemistry.* 1962, 7, 641-664.

Walkley, A. *Soil Sci.* 1947, 63, 251-254.

Wedepohl, K. H. *Handbook of geochemistry;* Springer: Berlin, 1979.

WHO (World Health Organization) *Environmental health criteria: mercury;* WHO: Geneva, 1976; 131 pp.

WHO (World Health Organization) *Guidelines for drinking-water quality;* WHO: Geneva, 1984; 128 pp.

WHO (World Health Organization) *Guidelines for drinking-water quality: mercury*; WHO: Geneva, 1993; 2nd edn., v. 1, p. 51.

Wikipedia 2009a Mercury (element) [http://en.wikipedia.org/wiki/Mercury_(element)]

Wikipedia 2009b Minamata disease [http://en.wikipedia.org/wiki/Minamata_disease]

Zobell, C. E. *Bull. AAPG.* 1946, 30, 477-509.

INDEX

E

F

G

H

I

Q

R

S

X

Y

Z